Recent Advances in Analysis, Design and Construction of Shell and Spatial Structures in the Asia-Pacific Region

T0172242

Recent Advances in Analysis, Design and Construction of Shell and Spatial Structures in the Asia-Pacific Region

Editors

Kok Keong Choong
School of Civil Engineering, Universiti Sains Malaysia,
Penang, Malaysia

Mustafasanie M. Yussof
School of Civil Engineering, Universiti Sains Malaysia,
Penang, Malaysia

Jat Yuen Richard Liew
Department of Civil and Environmental Engineering,
National University of Singapore, Singapore

CRC Press
Taylor & Francis Group
Boca Raton London New York Leiden

CRC Press is an imprint of the
Taylor & Francis Group, an **informa** business

A BALKEMA BOOK

CRC Press/Balkema is an imprint of the Taylor & Francis Group, an informa business

First issued in paperback 2021

© 2020 Taylor & Francis Group, London, UK

Typeset by Apex CoVantage, LLC

Library of Congress Cataloging-in-Publication Data
Applied for

Published by: CRC Press/Balkema
Schipholweg 107C, 2316 XC Leiden, The Netherlands
e-mail: Pub.NL@taylorandfrancis.com
www.crcpress.com – www.taylorandfrancis.com

ISBN: 978-0-367-24855-0 (hbk)
ISBN: 978-1-03-208230-1 (pbk)
ISBN: 978-0-429-28471-7 (eBook)

DOI: https://doi.org/10.1201/9780429284717

Contents

Figures

Tables

Contributors

Siu-Lai Chan is a professor at Department of Civil and Environmental Engineering, Hong Kong Polytechnic University, Hong Kong.

An-ying Chen is an associate professor at School of Civil and Hydraulic Engineering, Hefei University of Technology, China.

Khai Seng Chew is a graduate student at School of Civil Engineering, Universiti Sains Malaysia and an engineer at Sediabena Sdn Bhd, Malaysia.

Kok Keong Choong is an associate professor at School of Civil Engineering, Universiti Sains Malaysia, Penang, Malaysia

Seung-Deog Kim is a professor at Department of Architectural Engineering, Semyung University, South Korea.

Toshiaki Kimura is a lecturer at Nagoya City University, Japan (former staff of Sasaki Structural Consultant, Japan).

Jie Hu is an engineer at Institute of Architectural Exploratory and Design, Beijing University of Technology, China.

Shogo Inanaga is a graduate student at Department of Department of Architecture and Building Engineering, Tokyo Institute of Technology, Japan.

Nozomu Kogiso is a professor at Department of Aerospace Engineering, Osaka Prefecture University, Japan.

Xiongyan Li is a professor at Spatial Structures Research Center, Beijing University of Technology, China.

Jat Yuen Richard Liew is a professor at Department of Civil and Environmental Engineering, National University of Singapore, Singapore.

M. L. Liu is a senior engineer at Shanxi Provincial Academy of Building Research, China.

Renjie Liu is an associate professor at School of Civil Engineering, Yantai University, China.

S. W. Liu is with Alpha Consulting Limited, Hong Kong.

Y. P. Liu is with NIDA Technology Co., Ltd., Hong Kong.

Ryota Matsui is an associate professor at Hokkaido University, Japan.

Marijke Mollaert is a professor at Department of Architectural Engineering, Vrije Universiteit Brussel, Belgium.

Toku Nishimura is a professor at Department of Architectural Engineering, Kanazawa Institute of Technology, Japan.

Sun-Woo Park is a professor at Department of Architecture, Korea National University of Arts, South Korea.

Masao Saitoh is a professor emeritus at Nihon University, Japan.

Mutsuro Sasaki is a professor emeritus at Hosei University and president of Sasaki Structural Consultant, Japan.

Hin Cheong Saw is an associate principal at Arup Jururunding Sdn Bhd, Penang, Malaysia.

X. W. Song is an engineer at China Aerospace Academy of Architectural Design & Research Co., Ltd., China.

Toru Takeuchi is a professor at Department of Architecture and Building Engineering, Tokyo Institute of Technology, Japan.

Akira Tanaka is a lecturer at Department of Industrial Information, Tsukuba University of Technology, Japan.

Song Teik Tang is with Technical Department/Property North of SP Setia Sdn Bhd, Penang, Malaysia.

Yuki Terazawa is an assistant professor at Tokyo Institute of Technology, Japan.

Veng Wye Tong is a principal at Arup Jururunding Sdn Bhd, Penang, Malaysia.

Hai-ying Wan is a professor at School of Civil and Hydraulic Engineering, Hefei University, China.

Sik Kwang Wong is a principal at Arup Jururunding Sdn Bhd, Penang, Malaysia.

Jinzhi Wu is an associate professor at Spatial Structures Research Center, Beijing University of Technology, China.

Suduo Xue is a professor at Spatial Structures Research Centers, Beijing University of Technology, China.

Ji-po Yu is a graduate student at School of Civil and Hydraulic Engineering, Hefei University, China.

Y. G. Zhang is a professor at Spatial Structures Research Center, Beijing University of Technology, China.

Chapter 1

Design and construction of complex large-span structures

Jat Yeun Richard Liew

1 Introduction

This chapter presents the two recent large-scale complex structures in Singapore that are perceived to have significant impact in transforming Singapore into a global city in realizing architectural marvels and achievements. These complex steel structures required advanced analysis software and building an information model for pre-construction visualization, computer numerical control data for fabrication, and information sharing between designers and builders for construction. Because of their unique shapes and complex joint details, these structures were more difficult to fabricate and construct, and thus special efforts were needed to identify problems earlier and to design practical joint details for transportation and safe construction. Besides being grand, these projects were developed with thoughtful planning to merge with the immediate environments they inhabited. The projects presented herein also explore sustainable and green construction with an aim to lower carbon emissions and reduce their impact on biodiversity, to use less energy, water and other resources, and thus to minimize their impact on the built environment. This is made possible with the advancement in high strength and lightweight materials, design expertise, fabrication technology and erection technique, and, more importantly, the close cooperation and commitment of designers and contractors sharing knowledge and working as a team to ensure successful transformation of an innovative design to a successful completed project.

2 Gardens by the bay – the cool conservatory complex

The cool conservatory complex consists of a "cool moist" biome and a "cool dry" biome, as shown in Figure 1.1. The cool moist biome will re-create the cool and moist environment of a mountain "cloud forest," while the other will replicate the cool dry conditions of a Mediterranean spring. Research was done on energy modelling and state-of-the-art cooling technologies for the cool conservatory complex. The environment is controlled to accommodate special plants for winter/summer seasons which lead to varying internal temperature and humidity conditions.

Essential to the success of the planting strategy in the conservatory is the light transmission and heat gain through the envelope to the planted fields below. The structure is integrated with the façade system and shading system to minimize the overall silhouette that obstructs natural light entering the conservatory. This is achieved by using special triangular profiled sections for the gridshell to minimize obstruction of sunlight and automatic sunshade systems to adjust the amount of light entering the enclosure.

Figure 1.1 Flower Dome and Cloud Forest conservatories.

The plan dimension of the cool dry biome is 170 m × 105 m and the cool moist biome is 120 m × 85 m. The structural envelope of the biomes consists of a gridshell-arch steel structure with a double glazed skin that sits directly on the gridshell, as shown in Figure 1.2. The facade system is an integrated part of the steel arch and steel gridshell system and is considered as an integral component of the overall envelope. Conditions inside the cool dry greenhouse have to be maintained at 25°C, with a maximum relative humidity (RH) of 60% during the day. This will drop to 17°C at 65% RH at night. The night temperature may be lowered by 4°C to create the "end of winter" condition. In the cool moist greenhouse, the daytime climate requirements are daytime 25°C at 80% RH, dropping to 17°C at 80% RH at night. Again, the cooling system has the capacity to lower the night temperature by 4°C to simulate the end of winter. The challenge for the engineers was to maintain these conditions sustainably with minimum use of energy.

2.1 Hybrid structural form

The structural plan and elevation views of the Cool Dry Conservatory are shown in Figures 1.2a and b, respectively. The main structure consists of curved arches of varying cross section span from north to south supporting the gridshell structure below (see Figures 1.2a and b). The arches carry both axial forces and moments due to their essential function of resisting asymmetrical loading and stabilizing the gridshell against buckling. The arches are connected laterally by stainless steel rods which stabilize the arches against lateral buckling (see Figure 1.3). The gridshell comprises of linear elements which are triangular in shape. The elements resist principally axial loads but also resist in-plane moments (due to Vierendeel action) and out-of-plane moments (due to glazing loads, wind loads, maintenance loads, etc.). Varying triangular members link the arches and gridshell at regular intervals. The gridshell is supported and stabilized against buckling by the arches. Bracing elements run along the nose of the gridshell in the plane of the gridshell, and tension rods are used to support the gridshell from the arches, as shown in Figure 1.4. The gravity load on the

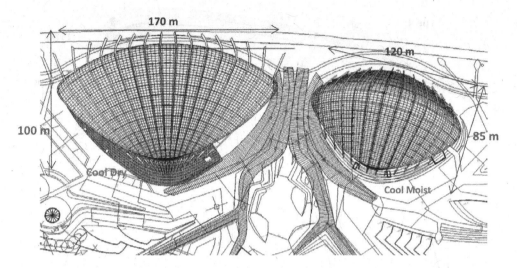

170 m

120 m

100 m

85 m

Cool Dry

Cool Moist

Figure 1.2a Geometry of the Flower Dome and the Cloud Forest – plan view.

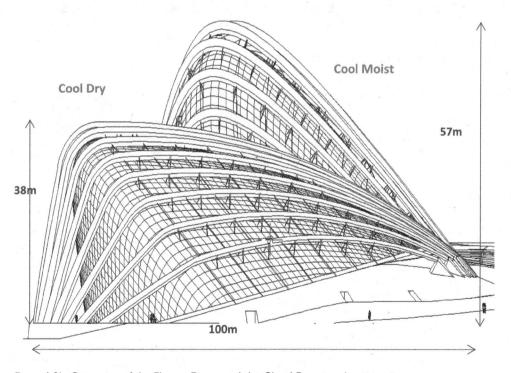

Cool Moist

Cool Dry

57m

38m

100m

Figure 1.2b Geometry of the Flower Dome and the Cloud Forest – elevation view.

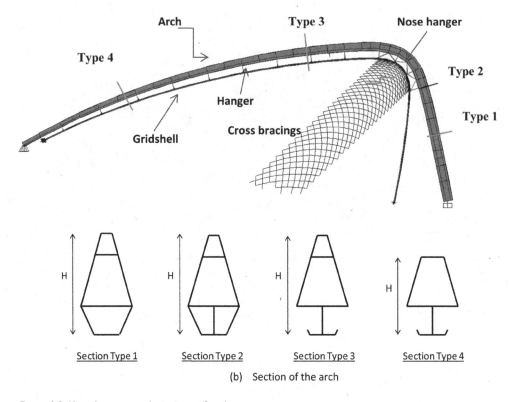

(b) Section of the arch

Figure 1.3 Key elements and sections of arches.

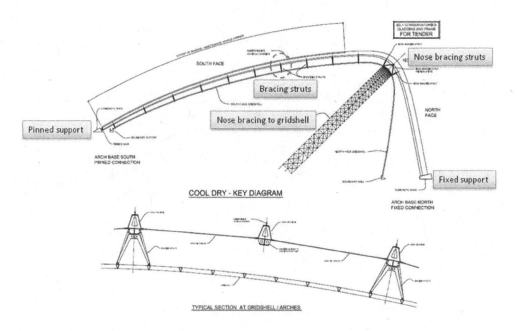

Figure 1.4 Key elements and boundary conditions.

gridshell is supported by the arches, whereas the lateral stability of the arches is provided by the gridshell. In this unique hybrid system, the gridshell and arches complement each other in resisting gravity and horizontal loads.

2.2 Gridshell members

The structural members of the gridshell are made from S355 steel plate cold formed to a triangular cross-sectional shape (width = 180 mm, depth = 230 mm) with one longitudinal weld seam as shown in Figure 1.5. The structural members were then hot-finished to reduce the influence of welding residual stresses. Since the modern code did not provide any guidance for designing triangular members, a test program was carried out to investigate the flexural and buckling resistance of the triangular gridshell members of various plate thicknesses (8, 10, and 12 mm).

The test setup for flexural tests and compression (buckling) tests are shown in Figures 1.6 and 1.7, respectively. The test results are compared with the code's predictions to ensure that the existing method of prediction can be applied for designing triangular shape structural elements.

The tests concluded that the gridshell members can develop full plastic moment capacity under flexural loading. The design moment resistance calculated based on the BS5950 code is conservative. The gridshell members behaved in a ductile manner under bending with no sign of cracking or fracture at the corner or at the weld seam of the cross sections. The hogging and sagging moment capacities of the section are almost identical. The axial buckling tests showed that BS5950:Part1:2000 may be used to predict the axial buckling capacity of gridshell members. The design axial buckling capacity calculated based on BS5950:Part1:2000 is conservative compared to the test results.

2.3 Advanced modelling and analysis

Three-dimensional nonlinear analysis was used to determine the force distributions in the structural members and the connections. The structural analyses were carried out using

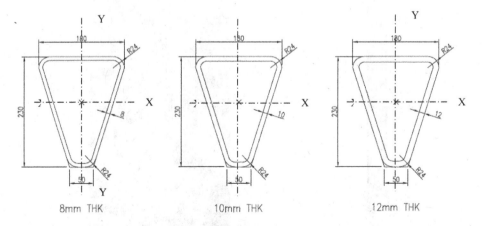

Figure 1.5 Gridshell members with triangular profiles.

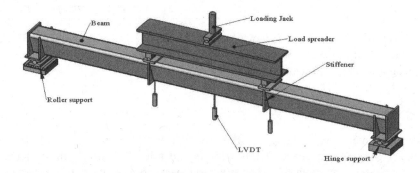

Figure 1.6 Test setup to determine flexural resistance.

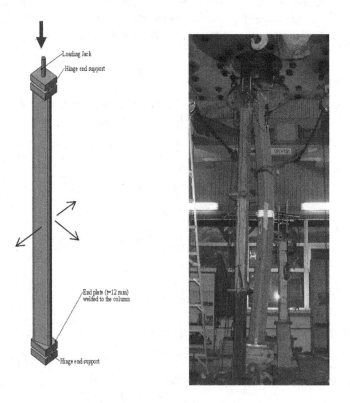

Figure 1.7 Test setup to determine buckling resistance.

second-order elastic analysis software, taking into consideration the following (Liew *et al.*, 1993; Liew and Tang, 2000; Liew and Chan, 2010):

*   Geometric nonlinear behaviour including global stability (P-Δ) and member stability (P-δ) effects. This included the effects of displacement and rotation on equilibrium and compatibility equations.
*   The assumption of material to be elastic up to the factored load level.
*   Sequence of construction (i.e., load sequencing).
*   Global imperfection (L/250) and member imperfection (l/300).
*   The assumption of rigid joint models, as fully welded joints were adopted in construction.
*   Wind, temperature, movable maintenance loads, support settlements, robustness, and fire.

The advantage of the direct second-order analysis method is that it includes member and global initial imperfection effects and thus it can capture the second-order forces directly from the analysis (Liew and Chen, 2004). This eliminates the potential error of using the conventional first-order linear elastic analysis followed by member capacity checks which require the estimation of member effective length. The first-order analysis with the moment amplification method given in the code cannot be used to predict the secondary forces because of the complex geometry of the structure. Moreover, first-order linear analysis is not recommended for designing three-dimensional structures of unusual shape and complex geometry. The structural model for second-order analysis and design considers all structural elements such as steel arches, gridshell, nose bracing, arch stability cables, and hangers, modelled by beam-column element with initial imperfections. The glass panels are ignored in the model, and the loads applied to the glass panels will be transferred to gridshell and arches by panel elements which do not have any stiffness.

The structural analyses follow the construction load sequences to identify the most critical stressed element for design consideration. In the construction, the steel arches were allowed to deflect before connecting them to the grid's structure. Otherwise, the self-weight of the arches will be transferred to the gridshell during construction, causing overload on the gridshell.

The end moment release of selected gridshell members adjacent to the nose bracing at the hangers' positions are required to reduce the stress due to the thermal expansion of members under the change of temperature. The end moment release was facilitated by adopting bolted (pin) end connection of these gridshell members. Figures 1.8a and b show the aerial view and the construction of the cool dry and cool moist conservatory complex. Figures 1.9a–d show the supports and internal view of the lattice shell for the cool dry conservatory. Figure 1.10 shows the internal views of the cool moist conservatory including the aerial foot bridge surrounding the forest mountain.

(a) Cool dry conservatory   (b) Connection between arches and gridshell

Figure 1.8 Views of cool dry and cool moist conservatories after construction.

(a) Fixed support for arch   (b) Pin-support for the arch

(a) lattice shell structure

(d) Close-up view of the gridshell

Figure 1.9 Various views of curved arches and gridshell for the cool dry conservatory.

Figure 1.10 Internal view of the cool moist conservatory.

3   Changi Jewel – a large-span glass roof with oculus

The Jewel project is a new retail and lifestyle complex at Changi Airport, Singapore. The complex consists of five basements and sits five storeys above ground. The multi-storey complex is enclosed in a distinctive glass-and-steel dome featuring a rain vortex waterfall from a height of 40 m within a Forest Valley with a terrace garden. The perspective view of the Jewel roof is shown in Figure 1.11. The top floor is a canopy park featuring bouncing and walking nets, a 165-foot sky bridge, two mazes, a giant slide, and eight bars and restaurants. The exterior of the 10-story building is made of glass and crisscrossed with an aluminum-and-steel framework.

The overall dimensions of the jewel roof are 200 m along the longitudinal axis and around 150 m along the transverse axis occupying a floor area of approximately 24,000 m², as shown in Figure 1.12. The maximum height of the roof is approximately 36 meter from the level 5 deck, as shown in Figure 1.13. The total steel tonnage for the roof is about 6000 tons.

The roof of the Jewel complex is made of single-layer diagrid steel comprising of 33 hoops and interconnected bias members supported by 14 tree columns and ring beam located on the fifth storey, as shown in Figure 1.14. The diagrid structure was formed by triangular grids with bias and hoop members bolted together via solid nodes, as shown in Figure 1.14.

Advanced analysis is carried out to ensure the overall stability of the roof structure under various load combinations, including accident limit states involving fire and blast scenarios.

(a)   Perspective view

(b) Glass and steel dome with water fall feature

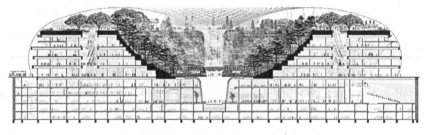

(c) Section view showing retail and garden

Figure 1.11 Perspective models of the Jewel roof.

Source: Courtesy of Jewel Changi Airport Development.

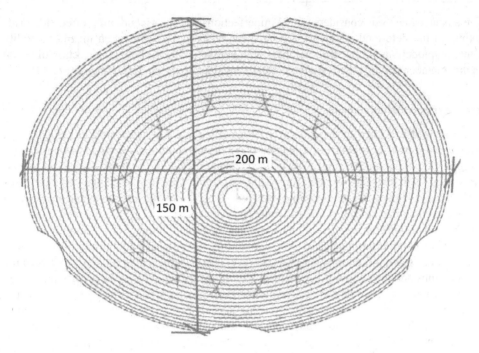

Figure 1.12 Plan view of the Jewel.

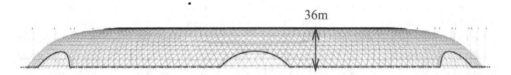

Figure 1.13 Elevation view of the Jewel.

Figure 1.14 Rectangular steel members joined by solid nodes to form triangular grids.

The advanced analysis considers global imperfections and the second-order effect due loads acting on the deformed geometry. The analysis takes into consideration member stability effect by modelling all the steel members with equivalent initial bow imperfection of L/300 so that member stability check is automatically included in the analysis (Chan *et al.*, 2017).

3.1   Structural concept

The steel roof is formed by a single-layer lattice dome made of hoop and bias elements that intersect to form triangular grids. The hoop elements are structurally arranged along circumferential directions and are made from welded rectangular hollow sections. They resist principally axial loads and also moments due to glazing loads, wind loads, and maintenance loads.

The bias elements are structurally arranged along the radial directions and made from welded rectangular hollow sections. Similarly, they resist both axial loads and moments. The bias and hoop elements form the skeleton of the roof structure. The entire roof could be divided into two parts according to force distribution: the part between the ring beam support and the tree columns forms arches to resist the external loads, whereas the part between the tree columns and the oculus utilizes tensile membrane action to resist the gravity loads. The hoops elements surrounding the tree columns form a compression ring so that they could balance the tensile membrane forces from the bias elements at the oculus, as illustrated in Figure 1.15. The compression ring elements and tensile membrane action from the oculus elements form a force-balancing mechanism to achieve an efficient structural system to accommodate an ocular near the middle of the roof structure.

3.2   Precision fabrication

The hoop and bias elements were rectangular hollow sections made from four S355 steel plates welded at the four corners. Since there were many members and much straight welding to be done to form the bias and hoop beams, robotic welding was adopted to fabricate the members to achieve high-quality workmanship, enhance productivity, reduce defects, and ensure good fitting at the site to form the desired geometry. High-quality precision fabrication was a key factor to ensure successful completion of this structure within a tight construction schedule.

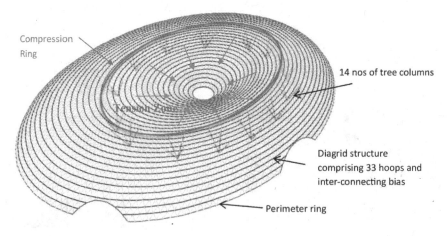

Figure 1.15 Diagrid roof and its supporting structures.

3.3 Beam to node connections and joint modelling

The bias and hoop beams were bolted to solid node connectors to form the skeleton of the roof structure. The solid steel node was machined from a very thick steel plate with threaded holes so that the members could be bolted to the node from the inside of the box sections. The connections between the solid nodes and the bias and hoop beams were semi-rigid and of partial strength. Preloaded high-tensile bolts were used to connect the members and the nodes. The typical semi-rigid joints between the solid steel node and the hoop and bias beams are shown in Figure 1.16. The hoop and bias beams were bolted into a solid steel node through the end plate which is welded inside the beam end with about 20 mm set back. The bolts were tightened from the top openings on the hoop and bias beams using a special calibrated torqued wrench to achieve the required pre-tensioning. The rotational stiffness of the semi-rigid joint was determined based on the section dimensions and the bolts sizes using the component method (EN1993–1–8, 2005).

The connection rotational stiffness S_j was calculated in accordance with EN 1993–1–8 and defined as the secant stiffness in the design moment-rotation characteristic as shown in Figure 1.17. In principle, the design rotational stiffness was determined based on the properties of its basic components.

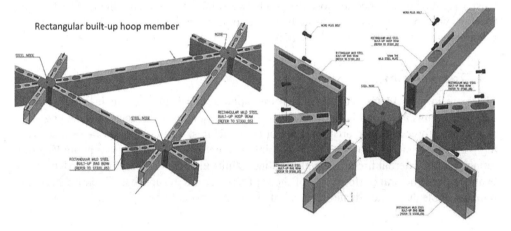

Figure 1.16 Typical semi-rigid joint between hoop and bias beams.

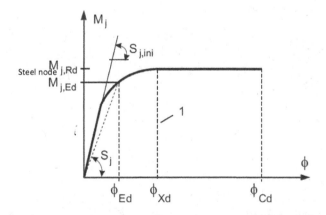

Figure 1.17 Typical design moment-rotation characteristic.

The connection between the solid steel node and the connecting member was modelled using a rotational spring element with spring stiffness, S_j, and moment resistance, $M_{j,Rd}$. Second-order semi-rigid analyses were carried out on the dome structure under various load combinations. The design moment and shear force at the beam's end were checked against the connection's moment and shear resistance to ensure that the bolts are adequate.

The basic component of the bottom flange and webs of the box beam in compression was not taken into account for the calculation of the rotational stiffness as recommended by EN 1993–1–8. The deformation of the basic component of the end plate in bending was not considered since the end plate was relatively thick and its deformation due to the bolt tension force is rather small. This assumption is valid as the end plate behaves like a one-way supported rigid plate, and the span of the one-way plate is rather short, equal to the box section width minus twice the web thickness and weld sizes. Therefore, the deflection of the end plate can be ignored in the stiffness calculation. The overall rotational stiffness of the semi-rigid joint was determined based on the axial stiffness of the bolts.

For a typical joint shown in Figure 1.18, the spring components representing the stiffness of the bolts for the calculation of joint rotational stiffness is illustrated. The contribution of the bolts' pretension forces to the joint's rotation stiffness was ignored. Nevertheless, pretensioning the bolts helped to close the gap between the beam end and the node. Thus, all the bolts were pre-tensioned to ensure that the gap between the beam end and the node was minimum under the service load condition.

3.4   Tree columns

There are in total 14 tree columns, each tree column consisting of four inclined circular tubular columns as shown in Figure 1.19. These tree columns are crucial in providing support to the aforementioned arch and membrane structures. The columns were infilled with high-strength cement grout with internal reinforcement bars to achieve two hours ISO fire rating without external fire protection. Nonlinear finite element analyses were carried out to evaluate the structural performance of the roof structure to ensure that the entire structure remains stable due to accidental removal of any of the tree columns.

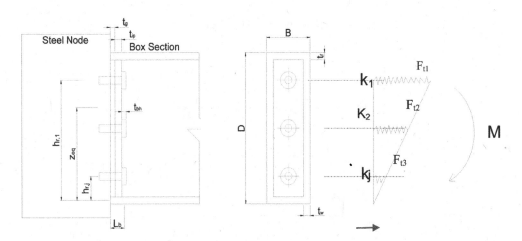

Figure 1.18 Semi-rigid joint model.

Figure 1.19 Tree columns consisting of a circular steel section infilled with high-strength reinforced concrete grout to achieve fire resistance.

3.5 Ring beams

The ring beams run along the outer perimeter of the roof and support the structure between the ring beam and the tree columns. The ring beams are made of welded box sections. The ring beam supports were modelled using semi-rigid linear springs with stiffness obtained from load displacement analysis of the supports.

3.6 Structural modelling and analysis

The structural analyses were carried out using second-order elastic analysis software, taking into consideration the following:

*   Geometric nonlinear behaviour including global stability and member stability effects. This includes the effects of displacement and rotation on the equilibrium and compatibility equations.
*   Material is assumed to be elastic up to the factored load level.

A three-dimensional analysis model is used to determine the forces in the members and connections. The structure is analysed by direct second-order elastic analysis. The advantage of this method is that it includes member and global initial imperfection effects and thus can capture the second-order forces directly from the analysis. This will eliminate the potential error of using the conventional first-order linear elastic analysis followed by member capacity checks which require the estimation of member effective length. The first-order analysis with the moment amplification method given in EN 1993–1–1 cannot be used to predict the secondary forces because the structure is rather sensitive to buckling with a very low elastic critical load factor, λ_{cr}. Moreover, first-order linear analysis is not recommended for used in dealing with three-dimensional structures of unusual shape or complex geometry.

3.6.1   Semi-rigid joint analysis

Nonlinear second-order semi-rigid analysis was used to analyse the structure in accordance with the Eurocode-3. The solid joints of the hoop and bias members were not explicitly modelled. Instead, the self-weight of the joints was considered and the rotational stiffness was applied at ends of the hoop and bias members, and the axial stiffness was assumed to be as infinite. The tree columns were modelled as pin-connected to the floor supports.

The glass panels were considered as thin concrete slabs and they were modelled by shell elements to simulate the limited lateral restraints to the hoop and bias members. Therefore, lateral torsional buckling of the deep rectangular steel members could be ignored.

The second-order elastic analysis considering both P-Δ and P-δ effects were used. Cross-sectional checks were carried out to ensure no plastic hinge was formed within the design load combinations.

Local plate buckling of hoop and bias members was considered by the effective cross-sectional area in accordance with EN 1993–1–1. Construction sequence was also modelled to ensure that the roof structure remained stable during the installation.

3.6.2   Member imperfections

As no structures are perfect and free from defects due to initial crookedness and residual stress, imperfections must be considered when using the direct second-order analysis in accordance with the modern code design procedure. For all the welded rectangular members, the equivalent member imperfection is thus taken as $l/200$, where l is the length of member considered. The direction of the member initial imperfection was assigned in accordance with the curvature of the member deriving from the end rotations of the member based on the mode shape from elastic critical load analysis.

3.6.3   Global imperfections

The global frame imperfection was considered for the influences of essential manufacturing tolerances that might affect the overall stability of the structure. Herein, the maximum magnitude of global imperfection was taken as L/250. The L is the distance between inflection points of the adjacent crest/trough of the mode shape. The mode shape was determined based on buckling analysis for each load combination applied, as illustrated in Figure 1.20.

The steel roof structure was analysed by second-order elastic analysis, taking into account the member and global imperfections. The ultimate limit state (ULS) was checked based on the second-order analysis and checking the cross-sectional resistances.

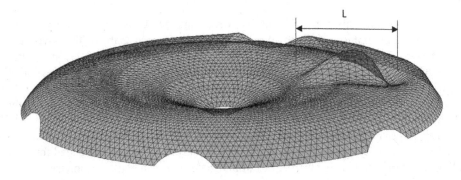

Figure 1.20 Determination of span length of global imperfection.

Critical members needed to be strengthened and replaced with larger sections according to the second-order direct analysis, with 265 load combinations for ultimate limited states design. These were compression members located around the oculus supported by the tree columns forming a compression ring.

It should be highlighted that the end stiffness of the hoop and bias members were the key parameters governing the structural resistance and integrity. The rotational stiffness of the node connectors was calculated based on Eurocode 3 1–8 component method, and their accuracies were verified by full-scale tests carried out on the node connectors and further confirmed by advanced finite element analysis.

3.7 Construction sequences

Because of the large size of the roof structures, the construction was carried out in stages under a fully propped condition. First, a crash deck was constructed between the oculus area and the tree columns. A working platform was erected about 1.5 m below the glass line as shown in Figure 1.21b. The structural members and nodes were assembled in this area. In stage 2 work, the structural members and nodes were assembled from the ring beam towards the column tree to form a segment of the roof (see Figure 1.21a). The entire roof structure was divided in 10 segments for the purpose of erection. The erection of steel structure started from Zone 1 and proceeded to Zone 10 in the clockwise direction until the structure was closed. Figure 1.22 shows the ocular roof and the overall structure during construction.

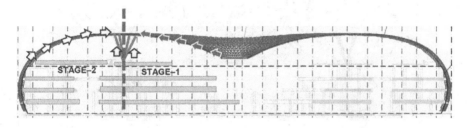

Construction sequence

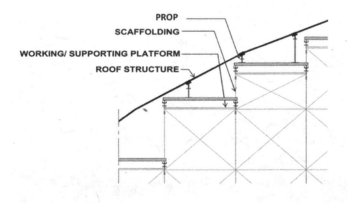

Propped construction

Figure 1.21 Construction sequence of the roof structures.

(a) Construction of the Ocular roof – insider view

(b) Construction of the Ocular roof – outside view

Figure 1.22 Construction of the roof structure.

4 Conclusions

This chapter presents two complex and large-span single-layer lattice steel roof structures that are perceived to have significant impact in transforming a city with modern facilities. Their complexity in terms of size, design, and craftsmanship were without equal in recent construction. These two structures faced similar challenges requiring a 3D building information model for pre-construction visualization, advanced modelling and analysis to ensure structural performance, computer numerical control data for fabrication, and information sharing between the designer and the builder for construction management. Because of their unique shapes and complex joints, these structures required high accuracy in pre-fabrication and site assembly, and thus special efforts were needed to identify problems earlier and to design practical joint details for efficient construction, transportation, and safe erection. These projects, besides being grand, were developed with thoughtful planning to merge with the immediate environments they inhabited. The key difference between the two projects was in the joints. The first involved the use of rigid and fully welded joints, and the second project adopted bolted node connectors. The speed of construction of the Jewel roof project was much faster but required high-precision fabrication to ensure all the structural members and nodes were accurately fitted together at the site. The roof structure for the conservatories has greater tolerance for construction, since small gaps between the node and connecting members were required for welding.

Steel is still the preferred material to produce innovative and spectacular architectural designs with complex geometrical profiles. This is made possible with the advancement in high-strength and high-performance material, advanced analysis design software, automatic fabrication technology and advanced erection equipment, and, most importantly, the close cooperation and commitment of designers, suppliers, fabricators, and erectors, sharing knowledge and working as a team to ensure successful transformation of an innovative design to a successful completed project of architectural and structural marvel.

Aware of global issues pertaining to climate change, the iconic structures described also explore sustainable and green construction with an aim to lower carbon emissions and reduce their impact on biodiversity, use less energy, water and other resources, and thus minimize their impact to the built environment. These structures are large in size, but what really makes them special are their uniqueness and the symbolic value when they were built, and the way in which they integrate the city's economic, cultural, and functional needs, making them truly iconic.

Acknowledgement

The author would like to acknowledge the published materials obtained from the following sources: Changi Airport Development, RSP Architects Planners & Engineers (Private) Ltd, CPG Consultants Pte Ltd, Safdie Architect, BuroHappold Engineering, Woh Hup (Private) Ltd, and Yongnam Engineering and Construction Pte Ltd.

References

Chan, S.L., Liu, Y.P. & Liu, S.W. (2017) A new codified design theory of second-order direct analysis for steel and composite Structures: From research to practice. *Structures*, 9, 105–111.

EN 1993–1–8, *Eurocode 3: Design of steel structures, Part 1–8: Design of joints*, European Committee for Standardisation.

Liew, J.Y.R. & Chan, S.L. (2010) *Second Order Analysis and Design of Cool Dry Conservatory at Gardens by the Bay Singapore*. Singapore: Special Report to Who Hup (Private) Limited.

Liew, J.Y.R. & Chen, H. (2004) Direct analysis for performance based design of steel and composite structures. *Progress in Structural Engineering & Materials*, 6(4), 213-228.

Liew, J.Y.R. & Tang, L.K. (2000) Advanced plastic hinge analysis for the design of tubular space frames. *Engineering Structures*, 22(7), 769–783.

Liew, J.Y.R., Punniyakotty, N.M. & Shanmugam, N.E. (1993) Advanced analysis and design of spatial structures. *Journal of Constructional Steel Research*, 42(1), 21–48.

Chapter 2

Iconic roof of Setia SPICE Convention Centre, Penang, Malaysia – architectural and engineering aspects of organic form single-layer latticed-grid steel shell roof

Hin Cheong Saw, Kok Keong Choong, Veng Wye Tong, Sik Kwang Wong, Khai Seng Chew, and Song Teik Tang

I   Introduction

The proposed lightweight iconic structure is a triangulated single-layer, latticed-grid, 400- to 600-mm-thick shell structure. It has a symmetrical kidney shape in the plan located at the main entrance. The longest and widest dimensions are approximately 113 m × 34 m. The height of the shell is around 24 m, which means it is a deep shell, and it is a very slender structure with longitudinal span to a minimum depth of element aspect ratio of 283.

The front half of the roof is covered by glass and the rear half by a composite panel. This wide-span roof frame is supported at four locations, i.e. two circular RC stairs at the rear and two RC columns at the front. The difference in level between the rear and front support is 15.0 m.

The shape of this shell is very organic and does not follow any one of the three typical uniform types, i.e., cylindrical, spherical, and paraboloidal shell (Figure 2.1a and 2.1b). In fact, it is a combination of these types: a cylindrical shell at the rear spanned in between two rear supports, a spherical shell at the front spanned in between the two front supports, and a paraboloidal shell in the middle. The stiffness of this iconic roof has vastly improved due to these combinations.

1.1   Architectural system

A large-span space truss can be divided into two main groups i.e. single-layer latticed grid and space grid (Figure 2.2a and 2.2b) (Choong et al., 2018). The choice will directly influence the overall cost, speed of construction, and aesthetic view. The factors influencing the decision are shape cum size of the roof plan, magnitude of loading, and support condition (Lan, 2000)

For a normal large-span roof, the architect is responsible for the overall architectural form and its functionality. The engineer, who understands load path, sizing of load-carrying elements, and control of the structural deformation within the acceptable limits for supporting roof finishes, is responsible for making it structurally safe. So, during the design process, the architectural shape of the structure is rarely questioned unless it threatens the structure suitability and constructability for the given purpose.

Figure 2.1a Isometric night view glass and composite panel installation.

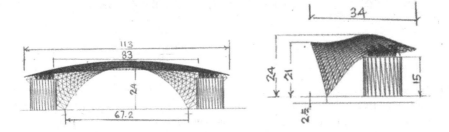

Figure 2.1b Front view and side view.

Figure 2.2a Spaced grid truss.

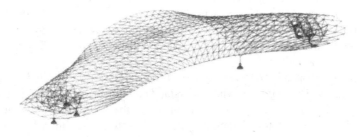

Figure 2.2b Single-layer latticed gridshell.

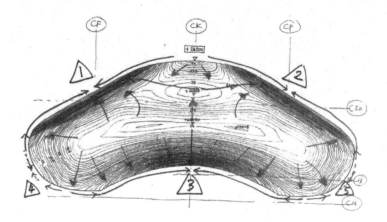

Figure 2.3 Roof contour and rainwater ponding check.

In contrast, for the design of a lightweight large-span shell roof structure, there is a need to understand the importance of deriving the stiffness of structure through changes in structure geometry, i.e., the rise, sag, and thickness of latticed shell which can be utilized to convert the loading resistance mechanism from bending to become more in tension and compression (Bangash and Bangash, 2003). Hence, full integration of both engineering and architectural skills are required. These confirm to a principal suites. "For an efficient structure use tension rather than compression or either in preference to bending."

Originally, the iconic roof has a more flatted top and a straight edge. This shape of structure called for a space truss with a spaced grids of 1.2 m deep in order to control the deformation to satisfy the limit of 1 in 360. However, the architect requested that the roof structure to be skinned structure consists of a single layer member. To achieve that structural geometry, the iconic roof is converted into organic form combining concave and convex shapes in order to increase its stiffness instead of the original shape of the more flatted top and straight edge.

By adopting this organic form, no rainwater ponding happens at the mid-span due to sagging deformation. Hence all rainwater will flow freely to the edge by gravity, as shown by arrows in Figure 2.3. As such, no rainwater gutter running below the roof is required, and a clean and neat ceiling soffit can be achieved.

1.2 Structural system

The design of single-layer gridshell structure follows a three-stage procedure.

1. The form finding
2. The grids patterning
3. The structural analysis

The form finding stage addresses the question of the surface geometry of shell spanning a given boundary configuration in which the stress distribution is tailored to be more membranal, i.e. in this system, the axial force is of paramount importance with as much reduction

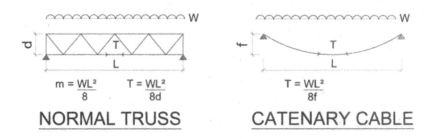

$$m = \frac{WL^2}{8} \qquad T = \frac{WL^2}{8d} \qquad\qquad T = \frac{WL^2}{8f}$$

NORMAL TRUSS   CATENARY CABLE

Figure 2.4 Analogy between normal truss and catenary cable.

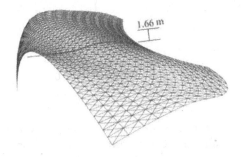

Figure 2.5 Initial model of iconic roof – difference in sag and rise is 1.66 m.

of bending stress as possible. The sag of the concave shape and the rise of the convex shape will become the depth of the space shell by which the stiffness of the space shell is derived from. This is an analogy to the catenary structure, as shown in Figure 2.4.

It can be seen that the tension in the bottom chord of the truss as well as in the catenary cable is actually derived by the same formula, $\frac{WL^2}{8d} = \frac{WL^2}{8f}$, where W is the uniform load, L is the span, d is the depth of truss, and f is the sag of the catenary cable.

From this analogy, the depth of a truss element, i.e., "d" by which the stiffness is derived, is equivalent to the sag of the catenary structure, i.e., "f."

So initially, the skinned roof is modelled as a more flatted top shell with shell geometry of 1.66 m, the intended load is applied, and huge deformation of more than 1 m is obtained (Figure 2.5). Then an improved model (Figure 2.6) is designed to adopt the deformed shape of the initial model, which is 3.1 m and reloaded to find a new state of equilibrium. This improved model has acquired extra stiffness due to the shell geometry of 3.1 m and hence further deforms only slightly and falls within the limit of 1 in 360. By adjusting the required member size and the element wall thickness, an aesthetically pleasing and cost-effective structure is obtained. This completes the form finding stage.

The grids patterning stage is to allow a three-dimensional shape of latticed shell to be translated into a two-dimensional cutting pattern to enable the manufacture of the shell.

The patterning system adopted for this iconic roof is a single-layer three-way grid frame with three directional members running vertically, horizontally, and diagonally to form the triangulated structural grid. The grid is made up of 600 triangles centred about the line of

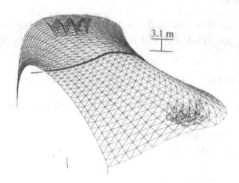

Figure 2.6 Improved model of iconic roof – difference in sag and rise is 3.1 m.

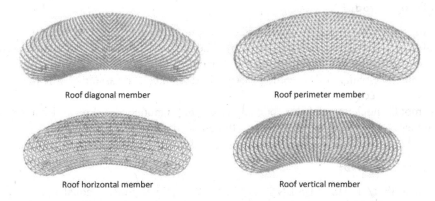

Roof diagonal member

Roof perimeter member

Roof horizontal member

Roof vertical member

Figure 2.7 Grid patterning for iconic roof.

symmetry, and so another 600 triangles of mirror shape, so each element has its own unique length (Figure 2.7).

For the analysis of space shell, the finite element method is by far the most accurate and effective. It can be used without any limit on the type, span, and shape of the structure, loading, and supporting conditions. For single-layer latticed shell, rigid joints have to be used. The key issue in the design of single-layer latticed gridshell is overall buckling (Gambhir, 2004).

2   Analysis and design criteria

2.1   *Design code & standard and coordinates system*

BS5950–1:2000 – Code of Practice for the design of rolled and welded section is adopted.
The structural axis system is as follows:

- +X horizontal from left to right
- +Y horizontal from bottom to top
- +Z vertically up

2.2 Computer software

The computer programs used in the analysis and design are Arup in-house Oasys General Structural Analysis Program GSA and STAAD.Pro V8i, of Bentley Systems.

2.3 Analysis parameter

The following properties are used for the analysis on the steel member:

Young's Modulus = 2.05×10^8 kN/m²
Shear Modulus = 7.885×10^7 kN/m²
Poisson Ratio = 0.3
Mass density of steel = 7850 kg/m³
Gravitational acceleration = 9.81 m/s²

2.4 Design model

The roof is modelled along the centre line of the steel section as a 3D latticed gridshell. The ends of the frame members are connected by rigid nodes. All the shell roof members are of a rectangular hollow section except for the support frames above the supporting stair core, where a mainly circular hollow section was used. All members are assumed to be frame type, i.e., rigidly connected.

The roof triangular area of glass or composite panel is modelled as 1 mm thick aluminium plate for ease of applying loading; i.e. as a three-noded plate.

2.5 Steel material

All CHS, RHS, and SHS to BS EN 10219 S275 (Ys = 275 N/mm²) or equal and steel plates and flat bar to BS EN 10025 S275 are used.

2.6 Design method

Two limit states analyses are adopted, i.e., the service limit state and the ultimate limit state.

The service limit states are related to performance under normal service conditions, e.g., deflections, and the ultimate limit states are related to safety and load-carrying capacity of the structure where the specified loads are multiplied by relevant load factor.

The design facility available in STAAD.Pro software is used to code check these structural members. Arup GSA Program is used to check the P-Delta overall buckling stability, which is the key issue in the design for single-layer latticed shell. A factor of 9 to 10 is called for the ultimate limit state load factor of 1.2 (SW + DL + LL + WP).

3 Loading assumption

The iconic structure is subjected to the following loadings:

1. Self weight of steel frame (SW), which is around 160 kg/m², automatically calculated by the program.

2. Dead load (DL) of glass panel = 40 kg/m², nodal weight of general E&L load (350 kg) at each node.
3. Imposed live load on roof for maintenance purpose = 0.25 kN/m²
4. 360° wind load as per wind tunnel test report

Due to the organic form of this iconic roof, it poses a more complex aerodynamic problem than the conventional portal frame roof system. Therefore, a small-scale model wind tunnel test was done at Monash University, Australia (Figure 2.8), in 2015. The wind loads coefficients are extracted from the wind tunnel test report (Bekele, 2015) which is done on 360 wind directions (Figure 2.9).

5. Uplift wind load of (WU) of 1 kPa and uniform wind pressure (WP) of 0.25 kPa.

Figure 2.8 Wind tunnel testing done in Monash University for aerodynamic issue.

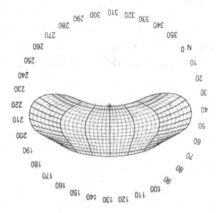

Figure 2.9 360° wind load.

4   Basic load case & load combination

The structure system is analyzed based on the following load cases and load combinations:

Load 1 (SW)   = Self weight (-Z) automatically calculated by the program
Load 2 (DL)   = Dead Load of roofing, nodal weight, general E&L
Load 3 (LL)   = Roof live load
Load X (WX) = Wind load from X deg where X is from 0° to 360°

4.1   Summary of applied load

This is shown in Table 2.1.

4.2   Service load combination

LOAD 101 = (SW + DL); LOAD 102 = (SW + DL) + LL; LOAD 10x = (SW + DL) + WX
LOAD 139 = (SW + DL) + LL + WP; LOAD 140 = LL + WP

4.3   Ultimate load combination

LOAD 201   = 1.4 (SW + DL); LOAD 202 = 1.4 (SW + DL) + 1.6LL
LOAD 20X = (SW + DL) + 1.4WX; LOAD 239 = 1.2 (SW + DL + WP)
LOAD 30X = SW + 1.4 WX; LOAD 339 = SW + 1.4 WU

5   Boundary condition

The Iconic Roof is supported at two nodes of the front column (nodes 1 and 2) spaced at 67.2 m apart, and at two pairs of supports (nodes 97 to 100) on top of two RC stair cores at 82.7 m and 84 m apart, respectively, as shown in Figure 2.10a. A spaced frame support system is formed above these two RC stair cores to strengthen the structure and control the cantilever deflection of the roof.

Due to the form of the roof, huge horizontal forces will be exerted on the supporting structures. To minimize or eliminate these forces, a pair of the support (nodes 99 and 100) on the RC stair core at the right are proposed to be free in X-horizontal direction, while the other pair of support (nodes 97 and 98) is always in pinned condition (Figure 2.10a). For all these supports, pot-type bearing is proposed (Figure 2.10b and 2.10c).

Table 2.1 Summary of Loads

LIC		Fx (kN)	Fy (kN)	Fz (kN)
1	Self Weight	0	0	−6340
2	Dead Load	0	0	−3950
3	Live Load	0	0	−1030
4	Wind X °	Max 63	Max 91	Max 659
5	Wind Pressure, 0.25 kPa	0	−186.837	−876.720
6	Wind Uplift 1.0 kPa	0	747.347	3510

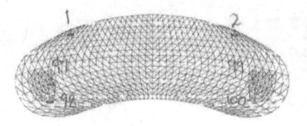

Figure 2.10a Six support locations are shown.

Figure 2.10b Sliding bearing pad at nodes 99 and 100.

Figure 2.10c Pinned bearing pad for other supports.

6   Analysis output

The design function available in the software STAAD.Pro is used to code check the member under various loading combinations while the software GSA is used to check buckling stability. Maximum member interaction ratio of 0.9, minimum buckling factor of 9.4, and natural frequency of 1.23 Hz are obtained. Based on the result of natural frequency of 1.23 Hz, it is concluded that the roof structure is not a dynamically sensitive structure (Figure 2.11).

The roof structure uses a rectangular hollow section, whereas spaceframe support members are mainly of a circular hollow section. There are 33 different sizes, and of these, 24 are RHS, 1 is SHS, and the remaining 8 are CHS. The majority of the member sizes used in

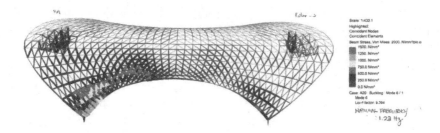

Figure 2.11 Buckling load factor of 9.4 and natural frequency of 1.23Hz.

RHS 400×200×6wt   RHS 600×400×50wt

Figure 2.12 Member sizes adopted.

this project are RHS $400 \times 200 \times 6$wt. Near the edge and supports, member sizes increase to maximum RHS $600 \times 800 \times 50$wt (Figure 2.12).

7   Node displacement

7.1   *Horizontal movement at sliding supports nodes 99 and 100*

The horizontal movement of the sliding support extracted from load case of self weight (SW) and dead load (DL) are summarized in Table 2.2.

Therefore, the total horizontal sliding movement in x direction $= 59 + 49 = 108$ mm

Hence, the pot bearing at nodes 99 and 100 is to be designed for minimum 200 mm lateral horizontal movement for load case SW and DL. After that, under LL and WP, the lateral reaction can be resisted by pot-bearing friction and hence no movement can be assumed.

7.2   *Node displacement summary*

The iconic roof has experienced the following net displacement for imposed live load and wind pressure:

Max X horizontal displacement at node 56 $= 202 - 178 = 24$ mm
Max Y horizontal displacement at node 46 $= 190 - 160 = 30$ mm
Max Z vertical displacement at node 514 $= 502 - 421 = 81$ mm

Table 2.2 Horizontal movement at supports 99 and 100

L/C	Node	X (mm)	Y (mm)	Z (mm)	Resultant (mm)	r_X (rad)	r_Y (rad)	r_Z (rad)
1:LOAD CASE 1-SELFWEIGHT	99	56.605	0.000	0.000	56.605	−0.002	−0.002	−0.000
	100	58.472	0.000	0.000	58.472	−0.001	−0.003	−0.001
2:LOAD CASE 2-DEAD LOAD	99	47.370	0.000	0.000	47.370	−0.002	−0.002	−0.000
	100	48.406	0.000	0.000	48.406	−0.001	−0.002	−0.000

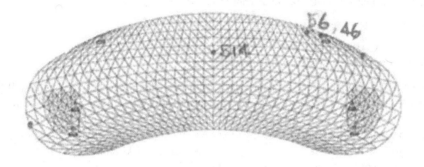

Figure 2.13 Nodes with max x, y, and z displacement.

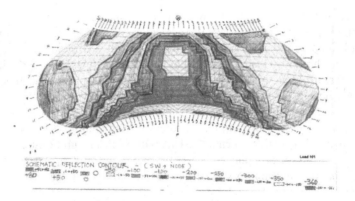

Figure 2.14 As-built deflection profile under self weight.

The deformation limits for LL and WP are checked as follows:

Allowable horizontal displacement = H/500 = 23,600/500 = 47 mm > 30 (OK)
Allowable vertical displacement = Span/360 = 67,200/360 = 186 mm > 81 (OK)

From this check, the net displacement is found to be within the allowable limit (Figure 2.13). Figure 2.14 shows the as-built deflection profile under self weight. It compares closely with the computer analysis output.

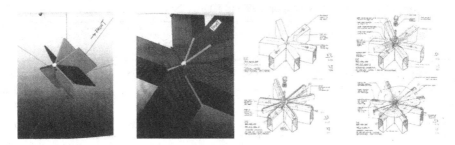

Figure 2.15 Cruciform fin plate roof node connection details.

8   Node connection details for fabrication requirements and quality control

The success of a shell structure depends totally in its connection design to achieve a full frame action so that the stiffness of the shell is not compromised as per the computer model requirement.

At every node, all together six members of hollow section from the patterning grids – i.e., vertical, horizontal, and diagonal – are met. In order to avoid loss in welding capacity caused by weld over weld in between members, a cruciform fin plate connection is proposed (Figure 2.15). This method is very suitable as for this fabrication, 100% of tensile testing results are passed.

9   Method of erection and fabrication sequence

Due to the three-dimensional organic form of this iconic shell structure and the introduction of sliding supports, all the elements have unique lengths and angles of connection. For easy fabrication, it is proposed to erect the shell in the air on full scaffolding (Figure 2.16). This will also allow the three-way load transfer behaviour to be achieved by simultaneous depropping of the full scaffolding. In order to obtain a symmetrical shape due to sliding supports, the fabrication sequence has to be planned properly (Figure 2.17).

After erection of the full shell structure, simultaneous depropping is a very important procedure. This is because loads can transfer from released temporary supports to adjacent temporary support if it is done sequentially. This will result in significant accumulation of load and cause catastrophic propping collapse by which a dynamic impact to the shell structure and the six permanent supports will occur and total failure will happen.

10   Conclusion

This chapter has described construction of the single-layer latticed steel shell roof as the main entrance statement of Setia SPICE Convention Centre, Penang, Malaysia. Due considerations have been given to both architectural and engineering aspects in order to erect such an aesthetically pleasing structure. This is done through:

a.   Respecting the architectural wish to have a skinned thick shell
b.   Adopting deeper structural geometry to increase the stiffness of the structure and use of the cruciform fin plate node for easy fabrication
c.   Careful planning of providing a full scaffold for the erection and simultaneous depropping of the scaffold for the completion of the shell.

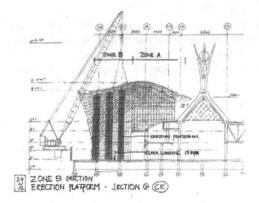

Figure 2.16 Full scaffolding temporary support.

Figure 2.17 Fabrication sequence.

References

Bangash, M.Y.H. & Bangash, T. (2003) *Elements of Spatial Structures: Analysis and Design*. Thomas, Telford, UK.

Bekele, S. (2015) *Wind Tunnel Testing and Report for Proposed Subterranean Penang International Convention Exhibition Center (SPICE) Penang, Malaysia*. Global Wind Technology Services Pty Ltd (GWTS), Melbourne, Australia.

Choong, K.K., Saw, H.C., Wong, S.K., Chew, K.S. & Tang, S.T. (2018) Double layer space frame for Setia SPICE Convention Centre. *Proceedings of the International Association for Bridge and*

Structural Engineering Conference: Engineering the Developing World, IABSE 2018, 25–27 April, Kuala Lumpur, Malaysia, pp. 587–592.

Gambhir, M.L. (2004) *Stability Analysis And Design of Structures*. New York: Springer.

Lan, T.T. (2000) Codes and standards for space frames in China. *Proceedings of the Sixth Asian Pacific Conference on Shell and Spatial Structure, Seoul, Korea*, 16–18 October 2000, pp. 3–10.

Oasys GSA, General Structure Analysis Program by Arup Oasys Ltd, London.

Chapter 3

Application and recent research on direct analysis with completed projects in Macau and Hong Kong

Siu-Lai Chan, Y. P. Liu, and S. W. Liu

1 Introduction

Long-span roof structures are widely constructed in stadiums, shopping malls, public halls, warehouses, airport terminal buildings, and so on as they provide large internal spaces without obstructing structural supports like columns. The commonly used typologies of roof structures include single-layer grids, double- and multi-layer grids, and their combinations. The long-span roofs generally experience large deflections and are sensitive to initial imperfections and patterned loads, especially for single-layer domes. The structural design complexity includes the fact that roof members cannot be clearly classified as beams or columns as can their counterparts in conventional building structures, but most steel design codes are for buildings made of columns and beams. Furthermore, the checking of snap-through or snap-back buckling is beyond the capacity of the traditional linear design method based on the effective length approach due to the complex buckling mode involving large deflections.

2 Review of previous work

The computer model using one element per member not only increases the numerical efficiency but also removes the difficulty in modelling the member's initial imperfections. Therefore, an advanced and robust element for large deflection and simulation of the initial curvature is crucial. Liew *et al.* (1997) laid a foundation for applied research in that direction. Liew and Tang (2000) further developed an efficient technique for practical DA of frames with a tubular cross section, and the work is significant in affecting the code development and research for long-span structures. Chan and Gu (2000) developed the stability function element with the explicit consideration of member imperfections, a closed-form solution that is suitable for simulating the extremely slender members. More recently, Liu *et al.* (2014) derived the curved Arbitrarily-Located-Hinge element with an objective to capture the highly inelastic behaviours along the member length. The element tapering is developed here and used to design a practical structure which seems to be new and original in structural design technology.

A tapering I-sections member is shown in Figure 3.1. To model the member with tapering sections, two methods are usually adopted as the approximated stepped elements and the tapering element approaches. The approximated stepped element method assumes the distribution of flexural rigidity along the member length as linear, parabolic, or cubic in analysis and assigned manually, and the errors can be large when using improper assumptions. This method is tedious and requires division of a member into many elements; otherwise the

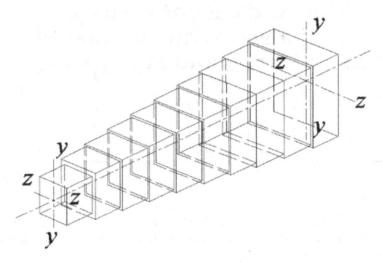

Figure 3.1 Stepped elements representation method.

deflections and stiffness can be wrongly estimated. The other modelling approach is the exact tapering element method. This method requires more element formulation efforts but reduces the computer efforts significantly. When using the approximate stepped element method, at least 20 elements are required when using this technique for an accurate simulation. In the present study, the extended tapered element model is used, and an exact analytical approach by explicit modelling of the non-prismatic members by the tapering stiffness factors is proposed.

In addition to element formulation, a kinematic method is required to describe the motion and large deflection of a deforming element. The incremental tangent stiffness method proposed by Chan (1992), is used and the equilibrium is established on the last configuration in the incremental-iterative procedure. This method is tested to be efficient and reliable.

3   Modelling of tapered I-section members

According to modern design codes such as de Normalización (2005), AISC (2016), and CoPHK (2011), the initial imperfections in both global frames and local members are the key factors which should be included in the DA process. The global and local imperfections can be considered by adjusting coordinates of each node and setting initial bowing of each member respectively based on eigen-buckling modes prior to DA. The former can be easily handled in the analysis procedures, while the latter should be supported by robust and reliable beam-column elements allowing for initial bowing such as the PEP element and the curved stability function (Zhou and Chan, 1995). As the non-prismatic member is required in this project, the "curved tapered-three-hinges" beam-column element (Liu *et al.*, 2012) allowing for member initial imperfection will be briefly introduced as follows for completeness. The local member basic forces and deformations of the element can be seen in Figure 3.2.

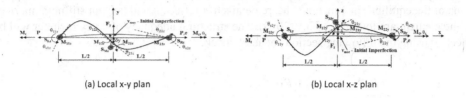

(a) Local x-y plan   (b) Local x-z plan

Figure 3.2 Forces versus displacements relations of the proposed element.

The Euler-Bernoulli hypothesis is adopted, and warping and shear deformation are neglected in the element. The applied loads are conservative and nodal. The total potential energy Π of the system can be expressed as the sum of the strain energy U and the external work done W as

$$\Pi = U - W \tag{1}$$

$$U = \frac{1}{2}\int_L EA\dot{u}^2 dx + \sum \frac{1}{2}\int_L EI_n(x)\ddot{v}_n^2\, dx + \sum \frac{1}{2}\int_L P\left(\dot{v}_n^2 + 2\dot{v}_{On}\dot{v}_n\right)dx + \sum \int_{\theta_{m,n}} S_{m,n}\theta d\theta\, (n=y,z) \tag{2}$$

$$W = \sum F_i u_i \tag{3}$$

in which EA is the axial stiffness constant and mean value for the non-prismatic section taken in this chapter; $EI_n(x)$ is the flexural stiffness about the y- or z-axis and changed with x along the member for the tapered section; $\theta_{m,n}$ and $S_{m,n}$ are the hinge rotations and stiffness of the internal plastic hinge at the y- or z-axis; P is axial force; u is the axial displacement function considering bowing effects; v_n is the transverse displacement function based on three-order polynomial; v_{On} is the transverse displacement function for initial imperfections. More details on the shape functions of u, v_n and v_{On} can be found in Liu *et al.* (2012).

By the minimum potential energy principle, the first variation of the potential energy function yields the equilibrium equations as:

$$\delta\Pi = \frac{\partial\Pi}{\partial u_i} + \frac{\partial\Pi}{\partial q}\frac{\partial q}{\partial u_i} = 0 \text{ And } i = 1\sim12 \tag{4}$$

The secant stiffness between forces and displacements can be obtained from Eq. (4). Further, the tangent stiffness matrix can be obtained by the second variation of the total potential energy function as:

$$\delta^2\Pi = \frac{\partial^2\Pi}{\partial u_i\partial u_j}\delta u_i\delta u_j = \left[\frac{\partial F_i}{\partial u_j} + \frac{\partial F_i}{\partial P}\frac{\partial P}{\partial u_j}\right]\delta u_i\delta u_j \text{ and } i, j = 1\sim12 \tag{5}$$

From Eq. (5), the tangent stiffness of the element can be formulated. In the incremental-iterative nonlinear analysis procedure, tangent stiffness is used as a predictor for estimating the displacement increment due to small but finite force, while the secant stiffness is used as a corrector of equilibrium. Generally, the Newton-Raphson method is used to find the solutions of the equilibrium equations. More details of the secant and tangent stiffness matrices of the element are found in NIDA (2013). The structural adequacy of every member will be checked using the following section capacity check:

$$
\frac{P}{P_y A} + \frac{\bar{M}_y + P\left(\Delta_y + \Delta_{0y}\right) + P\left(\delta_y + \delta_{0y}\right)}{M_{cy}}
$$

$$
+ \frac{\bar{M}_z + P\left(\Delta_z + \Delta_{0z}\right) + P\left(\delta_z + \delta_{0z}\right)}{M_{cz}} = \varphi \leq 1 \tag{6}
$$

where P is the axial force in member; A is the cross sectional area; P_y is the design strength; $\bar{M}_y$, $\bar{M}_z$ are the first-order bending moments about the minor and major axes; M_{cy}, M_{cz} are the moment capacities about the minor and major axes; Δ_y, Δ_z are the nodal displacements due to out-of-plumbness of frame sway induced by loads; Δ_{0y}, Δ_{0z} are the nodal displacements due to out-of-plumbness of frame imperfections; δ_y, δ_z are the member deformations due to loads on the member; δ_{0y}, δ_{0z} are the member deformations due to member initial bow; and φ is the section capacity factor. If $\varphi > 1$, a member fails in design strength check, and if $\varphi < 1$, the member section is very much overdesigned, and the member size can be reduced.

4   Verifications on the single element

Before the element is used, tests on the single element under various types of loading conditions are conducted to validate the element performance on element number convergence. A non-prismatic member with tapered I-sections is chosen for the present study. The overall width and depth of the member vary from 500 mm to 1000 mm and from 1000 mm to 500 mm, respectively. The plate thickness at the flange and web are 30 mm and 25 mm, respectively. The member length is 20 meters, and it is simply supported. The Young's modulus and Poisson's ratio are 205,000 MPa and 0.3, respectively. As aforementioned, the tapered sections stiffness factors can be calculated before the numerical incremental-iterative procedure.

The member will be analyzed by one of the proposed beam-column element under the following cases:

Case a: Pure bending about the major axis
Case b: Pure bending about the minor axis
Case c: Pure bending about both the axes
Case d: Uniaxial eccentric compression about the major axis
Case e: Uniaxial eccentric compression about the major axis
Case f: Bi-axial compression about both the axes

The conventional approach using stepped elements representation is adopted for the comparisons, with three types of modelling approaches selected as 5, 10, and 30 stepped elements per members. Herein, the analysis results from the model using 30 stepped elements are selected as the benchmarking solutions. The comparison results are plotted in Figures 3.3a–f.

From the comparisons, the proposed element improves efficiency and accuracy for a tapered beam-column. The results from the analysis model using one present element per member are closed to those from the model using 30 stepped elements, where the averaged difference is only 1.26%. However, the results from the models using 5 and 10 stepped elements per member are observed to have 18.56% and 5.34% discrepancy with the benchmark solutions.

As illustrated in the load versus deflections curves in Figure 3.3, the member deflections under different load conditions can be traced and predicted very well by the model using

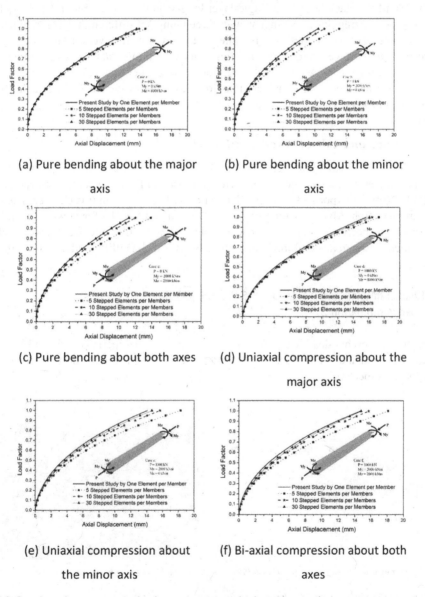

Figure 3.3 Benchmark test on a single element versus multiple prismatic elements.

only one proposed beam-column element. The example indicates that the numerical inefficient conventional stepped element representation can be replaced by the present element with numerical efficiency dramatically improved.

5 Design of MGM Spectacle Roof structure by DA

The proposed DA method equipped with the described element has been successfully applied to the design of the atrium "Spectacle Roof" at the MGM Resort in Cotai of Macau, which will be detailed herein as a real case study.

The MGM Spectacle Roof is a single-layer long-span thin shell structure, seen in Figure 3.4. This roof is irregular in geometry, with three humps of different heights, and takes the form of three vaulted "domes" with "valleys" in between. It covers a large public space between the hotel towers and associated function and gaming halls. All structural slopes of the roof are less than 30° and the rise-span ratio is not greater than 0.2.

In this section, the major considerations for the design of the MGM Spectacle Roof by DA will be introduced.

Design loads and critical load combinations

The design of the roof is in accordance with Macau codes for design loads and CoPHK (2011) for other parameters such as imperfections used in DA. The computer program used is NIDA (2013) incorporated with the element described earlier in this chapter. The design wind pressure from Macau loading code is 3.795 kPa and 3.3 kPa for downward and uplift wind cases, respectively. For this nonaccessible roof with fully enclosed glass panels, the most critical design load combinations are the downward and uplift wind over the entire roof surface, i.e.:

Load combination 1: "1.35 D + 1.5 Wd + 1.05 L"
Load combination 2: "1.0 D + 1.5 Wu"

where "D" represents dead loads including self-weight and superimposed dead loads; "L" denotes live load and is taken as 1.0 kPa; and "Wd" and "Wu" are the downward and uplift wind loads, respectively.

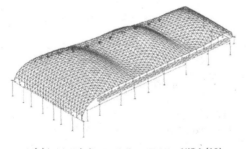

(a) Isometric in computer program NIDA (12)

(b) The space frame under construction

Figure 3.4 MGM Spectacle Roof under construction.

Although the DA is regarded as an innovative design method, like any other computer analysis method, a reliable design in terms of structural safety and cost-efficiency should be guaranteed only by correct input of design loads.

6   Imposing initial imperfections

No structures are considered as perfect and free from imperfections due to fabrication, transportation, erection, and so on. Two sources of imperfections should be included in the analysis process by DA, namely the member imperfection at the local element level and the frame imperfection at the global level.

Member imperfection may increase the bowing effect and induce the P-δ moment, which is important when the member is subjected to a large compression. Member imperfections include initial geometric imperfections and residual stresses. The initial geometric imperfections may be due to one or several aspects such as cambering, sweeping, twist, out of straightness, and cross-section distortion. The residual stresses can be due to manufacturing and fabrication processes. In de Normalización (2005) and CoPHK (2011), the two kinds of imperfections are simply combined into an equivalent geometric imperfection, while AISC (2016) adopts 0.1% member length as geometric imperfection and explicitly considers residual stress by a modified modulus approach. In this chapter, the equivalent initial bowing imperfections following CoPHK (2011) is adopted for the tapered I-beams. The beam-column element described in the previous section is used to model the tapered I-beams.

The frame imperfections are caused mainly by the out-of-plumbness of frame and column in the erection processes and during the construction stage. This type of imperfection may increase the sway effect and induce P-Δ moments, especially when the structure is subjected to large vertical loads.

For regular building structure, the notional horizontal force method can be used to consider the global imperfection. However, this approach is difficult to apply for long-span structures which do not have regular dimensions for the definition of height. An alternative and more reliable method for the complex structure is to use the elastic buckling mode as the imperfection mode, with an amplitude equal to height or span/300. The global imperfection is imposed by modifying the node coordinates, while the member imperfections can be enforced by setting the initial bowing to the same direction as the buckling mode shape.

The original design was based on the effective length method, which consumed some 20% more steel material measured in weight. This indicates the conventional effective length method could overdesign most members but underdesign some critical ones, making the structure heavier with a lower factor of safety.

7   Analysis and design outcome

The structure was completed, and it is now in use. In August 2017, Typhoon Hato visited Macau and caused heavy damage to various buildings, but not to the Spectacle Roof, which was noticed to have withstood the wind load safely.

In practical application for this project, the load sequence and offload process are also studied.

8 Acknowledgement

The authors are grateful to the financial support by the Research Grant Council of the Hong Kong SAR Government on the projects "Joint-based second order direct analysis for domed structures allowing for finite joint stiffness (PolyU 152039/18E)".

The authors would like to also acknowledge the design by the structural consultant Siu Yin Wai and Associates, who provided the supportive design for the project.

References

AISC, A.J.I. (2016) *Steel Construction Manual: American Institute of Steel Construction.* American Institute of Steel Construction, Chicago.

Chan, S.L. (1992) Large deflection kinematic formulations for three-dimensional framed structures. *Computer Methods in Applied Mechanics Engineering*, 95(1), 17–36.

Chan, S.-L. & Gu, J.-X. (2000) Exact tangent stiffness for imperfect beam-column members. *Journal of Structural Engineering*, 126(9), 1094–1102.

CoPHK (2011) *Code of Practice for the Structural Use of Steel 2011.* Buildings Department, Hong Kong SAR Government.

De Normalización, C.E. (2005) *EN 1993–1–1: Eurocode 3: Design of Steel Structures, Part 1–1: General Rules and Rules for Buildings.* Comité Europeo de Normalización, Brussels, Belgium.

Liew, J.R. & Tang, L.J. (2000) Advanced plastic hinge analysis for the design of tubular space frames. *Engineering Structures*, 22(7), 769–783.

Liew, J.R., Punniyakotty, N. & Shanmugam, N. (1997) Advanced analysis and design of spatial structures. *Journal of Constructional Steel Research*, 42(1), 21–48.

Liu, S.-W., Liu, Y.-P. & Chan, S.-L. (2012) Advanced analysis of hybrid steel and concrete frames, Part 2: Refined plastic hinge and advanced analysis. *Journal of Construction Steel Research*, 70, 337–349.

Liu, S.-W., Liu, Y.-P. & Chan, S.-L. (2014) Direct analysis by an arbitrarily-located-plastic-hinge element, Part 1: Planar analysis. *Journal of Constructional Steel Research*, 103, 303–315.

NIDA (2013). *Nonlinear Integrated Design and Analysis*, user's manual.

Zhou, Z.-H. & Chan, S.-L.(1995) Self-equilibrating element for second-order analysis of semirigid jointed frames. *Journal of Engineering Mechanics*, 121(8), 896–902.

Chapter 4

Structural design of free-curved RC surface – new crematorium in Kawaguchi

Toshiaki Kimura and Mutsuro Sasaki

I Introduction

The building presented in this chapter is the crematorium in Kawaguchi city, Saitama Prefecture, Japan. The architectural design process was started in 2011, and construction was completed in 2018. Toyo Ito Associates and Architects was responsible for architectural design, and Sasaki Structural Consultants took part in the structural design. This building is planned as a part of a park which is covered with plants. Figures 4.1 and 4.2 represent the status of this building in April 2018. The building's area is 5590 m², and its height is 13 m. Figure 4.3 shows the first-floor plan; cremation furnaces are located at the central position, and various areas for cremation are arranged around these spaces. Other space is open to the outside, so a visitor can see green in waiting rooms and the foyer, as shown in Figure 4.4. Free-curved RC surface is designed to be in harmony with the surroundings of this building. Its undulations remain steady on the north side. On the other hand, the central part of this roof fluctuates drastically. The plantation is arranged above the roof.

With respect to the structural planning and design, a method of computational morphogenesis is utilized. Through the consultation of many studies, the well-balanced configuration of the free-curved surface between the image of the architect and the structural rationality has been obtained. In addition, the total amount of self-weight can be reduced compared with conventional RC flat slab with the same span. This chapter also shows the construction process with a series of pictures showing various steps of construction.

In formwork construction, Rhinoceros and Grasshopper are utilized for construction drawing and checking the details. Through these digital tools, highly precise fabrication and installation of formwork elements can be realized.

The previous paper (Kimura *et al.*, 2016) reported the structural design, especially structural planning. The next report (Kimura & Sasaki, 2018) showed the abstract of the construction process. This paper represents an overall process of the structural design of the free-curved RC surface with detail of structural design and construction.

Figure 4.1 Exterior view of the front side.

Figure 4.2 Exterior view of the north side.

1:Entrance
2:Hall
3:Farewell room
4:Preparation room
5:Waiting room
6:Multti-purpose room
7:Shop
8:Nursing room
9:Rest room
10:Office
11:Furnace

Figure 4.3 First-floor plan.

Source: Credit by Toyo Ito Associates and Architects.

Figure 4.4 Interior view

Source: Photo credit by Toyo Ito Associates and Architects.

2 Structural design

2.1 Overall

To achieve architectural requirements, we propose the structural system, which consists of three elements (the free-curved RC surface, the RC rigid frame with the shear wall, and the steel column). The structural diagram is shown in Figure 4.5. RC shear walls are placed on the centre and the north part, and the steel columns are covered with concrete and arranged around the RC frames. Pile structure is used for the foundation.

2.2 Free-curved RC surface

With respect to the free-curved RC surface, a method of computational morphogenesis for the shell structures is utilized where the shape finding problem has been formulated as strain energy minimization problem. By using the proposed method (Kimura & Ohmori, 2018), the shape and the distribution of the thickness can be obtained easily. However, for considering the simplicity of the construction of the RC roof, the distribution of the thickness is set constant (200 mm) during the optimization process, and only the z-coordinates of the NURBS control points for the shape are adopted as the design variables. Control net and FEM mesh of the initial shape are shown in Figure 4.6a and 4.6b. The history of the optimization process is shown in Figure 4.6c. Black and white points represent the total strain energy and the maximum vertical displacement, respectively. The shapes of the free-curved surface at iteration are shown in Figures 4.7a–d. The colour shows the distribution of the

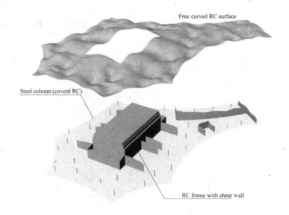

Free curved RC surface

Steel column (coverd RC)

RC frame with shear wall

Figure 4.5 Structural diagram.

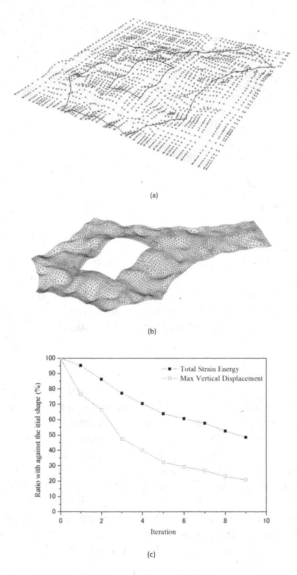

(a)

(b)

(c)

Figure 4.6 Numerical model and result, (a) control net, (b) FEM mesh, and (c) history of strain energy and maximum vertical displacement.

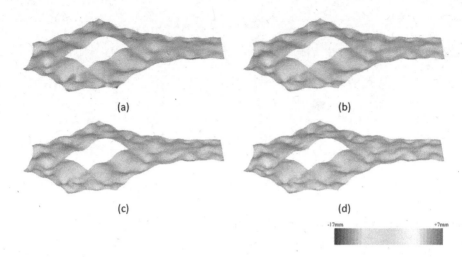

Figure 4.7 Shapes of the free-curved surface with the distribution of vertical displacement, (a) Initial shape, (b) Step 4, (c) Step 7, and (d) Step 9.

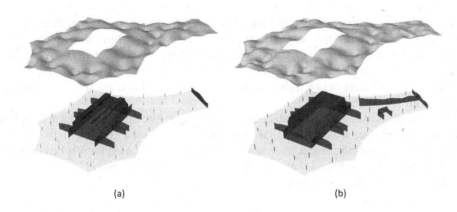

Figure 4.8 Examples of the configuration of free-curved RC surface, (a) SD phase, and (b) DD phase.

vertical displacement, and a blue colour (or red colour) represents large deflection. In the design process, the architect's aesthetics and structural rationality were satisfied at the shape of the free-curved surface of the ninth iteration shown in Figure 4.7d.

It should be noted that there were some studies between the architects and structural engineer, configurations of free-curved surface are shown in Figures 4.8a and 4.8b. The left figure represents the shape at the Schematic Design phase (SD phase), while the right figure shows at the Design Development phase (DD phase). As can be seen in Figure 4.8b, the RC wall is arranged at the north side. This wall is assigned to satisfy the architectural request to make the undulation of the roof on the north side lower to be in harmony with the surroundings.

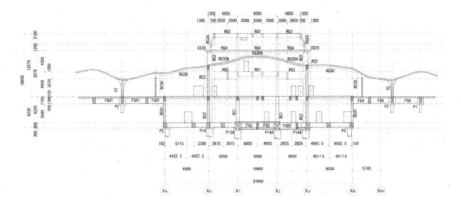

Figure 4.9 Structural drawing of elevation at Y1 line.

2.3   Supporting structure of free-curved RC surface (RC rigid frame with the shear wall, steel columns, and foundation)

The RC rigid frame with the shear wall supports the free-curved RC surface. RC shear walls are placed on the centre and the north part. These walls are 200 to 400 mm thick and are provided to withstand not only an earthquake but also any thrust force due to the vertical load. To support the free-curved RC surface appropriately, the steel columns covered with concrete are arranged around the RC frames. The stud bolt connects with the steel column for transmitting of stress between the RC surface and the steel column. This connection is covered with stiff RC element.

Pile structure is used for the foundation of this building. Pile tips are located at the gravel and coarse sand layer which is laid GL-54m under. Prestressed High-strength Concrete (PHC) pile is used mainly. Steel Concrete composite (SC) pile is used for pile head for seismic resistance and some RC piles with a device for collecting underground heat are provided to reduce the environmental impact.

3   Construction

3.1   Overall

Construction of this building started in January 2016. The construction process is shown in Figure 4.10. This chapter shows the construction process with a series of pictures showing various steps of construction. Particularly, we focus on formwork, rebar, and concrete construction.

3.2   Formwork construction

The formwork consists of the CNC girder, joist, and form board. Furthermore, curved form board which is reinforced by wooden ribs is utilized around the steel column and the edge of the surface. In this project, Rhinoceros is utilized for 3D modelling of formwork for sharing

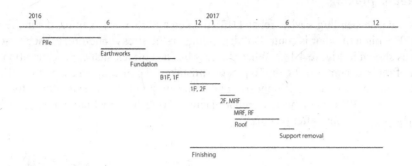

Figure 4.10 Construction process.

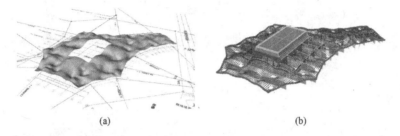

(a)   (b)

Figure 4.11 Isometric view of the 3D model: (a) design model, (b) construction model for formwork.
Source: Credit by KIYAMA Co., Ltd.

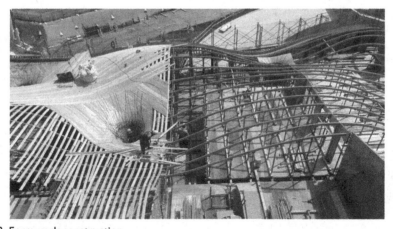

Figure 4.12 Formwork construction.

information between contractor and designer. Figures 4.11a and 4.11b show the design model named "Base model" (Hayashi *et al.*, 2019) and the formwork model, respectively.

Due to using these tools, the 3D model can validate the complex connection of RC free-curved surface, which cannot be understood easily, and data can be shared with the contractor appropriately. Highly precise fabrication and installation of formwork elements can be realized (Figure 4.12).

3.3 Rebar placing

To make elastic bending easily, rebar D13 (diameter = 13 mm) is utilized for the whole surface. The pitch of rebar is controlled depending on the stress condition of the free-curved surface as shown in Figure 4.14a. When the roof has a complex shape, the estimation of the amount of rebar becomes difficult. To prevent mistakes, we pick up an estimate of the rebar amount numerically by using Rhinoceros and confirming with the constructor before installation. Figure 4.13 shows the view of rebar placing during construction. Figure 4.14b shows the arrangement of rebar after placing.

Figure 4.13 Rebar placing.

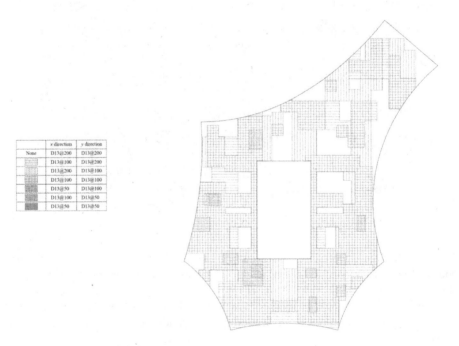

	x direction	y direction
None	D13@200	D13@200
	D13@100	D13@200
	D13@200	D13@100
	D13@100	D13@100
	D13@50	D13@100
	D13@100	D13@50
	D13@50	D13@50

Figure 4.14a Arrangement of rebar (lower layer).

Figure 4.14b Arrangement of rebar after placing (lower layer).

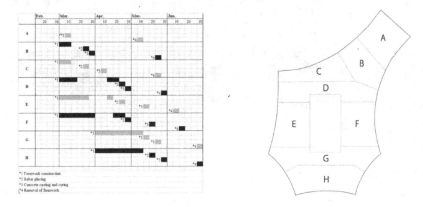

Figure 4.15 Construction process and layout of workspace for concrete casting.

3.4   Concrete casting

Generally, it is difficult to place concrete all at once on the whole surface because this roof has a large area. To solve this difficulty, workspace for concrete casting is divided into eight parts considering stress condition, and concrete placement is repeated (Figure 4.15). To build appropriately, a full-size mockup model is constructed before installation. We check and discuss the step of concrete construction through the mockup model (Figure 4.16). It should be pointed out that the installation of the mockup model is effective in realizing such a complex shape. Figure 4.17 shows the view during concrete casting.

Figure 4.16 Full-size mockup model.

Figure 4.17 Concrete casting.

4   Summary

This chapter reported the realization of a crematorium which has a free-curved RC surface. The detail of the structural design and construction process are represented. This chapter can be summarized as follows:

1. An overview of this building with respect to architectural design was presented, as well as an explanation of the structural system which is proposed to satisfy the architectural requirement.
2. Regarding the application of the method of computational morphogenesis to the RC free-curved surface, it should be noted that some studies between the architects and

structural engineer, for example, matched the architectural requirement and the structural necessity with the arrangement of the support of the free-curved surface.

3. Regarding the construction, various steps of the construction process (formwork erection, rebar placing, and concrete casting) are shown. It should be noted that highly precise fabrication and installation of formwork elements can be realized by digital tools, and the installation of a mockup model is effective in realizing such a complex shape.

Acknowledgments

We would like to thank Toyo Ito Associates and Architects for providing us with the building and other information, and to express special thanks to everyone involved in the design and construction of this project.

References

Hayashi, S., Yamasaki, K., Kimura, T. & Gondo, T. (2019) Construction process and rationalization of reinforced concrete roof with free-form surface by using NURBS model. *AIJ J. Techno. Des.*, 25(60, June), 941–946 (In Japanese).

Kimura, T. & Ohmori, H. (2008) Computational morphogenesis of free form shells. *J. Int. Assoc. Shell. Spatial Struct.*, Vol. 49(3), pp. 175–180.

Kimura, T. & Sasaki, M. (2018) Structural design of free-curved RC surface. *Proc. 12th Asian Pacific Conference on Shell and Spatial Structures (APCS2018), Penang, Malaysia.*

Kimura, T., Hiraiwa, Y. & Sasaki, M. (2016) Structural design of the crematorium in Kawaguchi. *Proc. of IASS Symposium 2016, Tokyo, Japan, Int. Assoc. for Shell and Spatial Struct.*, p. 62.

Chapter 5

Stability analysis and monitoring for the construction of a steel spatial hybrid structure stadium

Hai-ying Wan, An-ying Chen, and Ji-po Yu

I Introduction

The modern steel structure system is becoming increasingly complex, and the structure size is getting larger and larger. The stability of the structure during the construction directly affects the safety of the structure and of construction personnel, and it is also a matter of success or failure of the project. If structural instability occurs during the construction of large complex steel structures due to improper consideration, there will be catastrophic loss of life and property. The instability of the structure as a whole refers to the large geometric deformation or displacement of the structure caused by the deviation of most areas or almost the whole structure from the initial equilibrium position under the load. Structural stability is closely related to structural type, load form, and initial defects, and the analysis method is different from the stable analysis of the single member (Gao et al., 2012; Jing et al., 2008; Yuan et al., 2009; Borri & Spinelli, 1988). In addition to the analysis of the overall stability of the construction process, in order to ensure construction safety and construction quality, and at the same time to make sure the structure reaches or approaches the design state, it is usually necessary to use reasonable technical means to monitor the entire process of steel construction.

The large-span steel structure of a stadium (Figure 5.1) with a span of 107 m is comprised of prestressed cable, a space grid truss for a welding hollow ball, and a steel tubular truss. The steel structure consists of two parts, the inner ring and the outer ring. The inner ring is a mixed structure composed of a space grid truss for welding the hollow ball and its upper prestressed cable. The outer ring structure is a curved grid structure formed by 36 trusses of steel tubular truss and a brace system placed crosswise. The roof is covered with light steel purlins and aluminum alloy plates. The diameter of the support is 89.8 m, the plane projection diameter of the steel structure is 107.0 m, the height of the structure is 28.2 m, the projected area of the steel structure roof is 8392 m², and the total weight is about 1600 tons. The structure is shown in Figure 5.2.

The construction technique is introduced in the context of a stadium project, and the stability of incomplete time-varying structure is analyzed using MIDAS/Gen. We can use the results to guide structural construction and evaluate the structural safety. Based on the calculation and analysis, the monitoring technology and results of the structure are also analyzed.

Figure 5.1 Architectural effect drawing for the stadium.

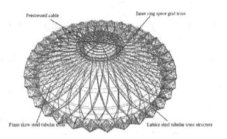

Figure 5.2 Steel structure axonometric drawing of the stadium.

2   Construction technique of the steel spatial hybrid structure

The construction of the steel spatial hybrid structure stadium started after the stadium grandstand, steel structural support columns, and embedded inserts were completed. The construction process is divided into three main construction stages, namely the installation construction stage, the unloading construction stage, and the prestressed cable tension stage.

The separate hoisting method was used in the installation construction stage. All the structural members were sent to the site in pieces and assembled into a hoisting unit on the site. First, 36 outer-loop supports were set up on the concrete structure platform. Then a crawler crane was adopted to complete the hoisting and installation of the truss. Finally, a lattice steel tubular truss structure was formed. The outermost ball joints fall on the outer-loop supports (Figure 5.3). Secondly, the erection of 35 middle-loop supports and completion of the corresponding 35 span truss hoisting and installation at the same time (Figure 5.4) were carried out, with one span truss and the corresponding middle-loop support at the outlet of the span site reserved for later hoisting and installation. Then eight inner-loop supports were set up. There were two support points above each inner-loop support, and the inner-loop support was distributed directly below the position of the inner-loop grid of the steel structure. At the same time, a central support was arranged in the central position of the stadium, to be used for the later installation of prestressed cable. When the inner-loop grid is assembled on the ground, it was cut into four symmetrical parts, then hoisted to the inner-loop support, respectively, and welded into the whole grid (Figure 5.5). Finally, the hoisting and installation of the remaining part of the truss, high-altitude assembly of suspended dome structure,

Figure 5.3 Outer-loop supports.

Figure 5.4 Middle-loop supports.

Figure 5.5 Installation construction of the inner-loop grid.

flying column, and prestressed cables were completed successively (Figure 5.6). After the completion of construction, the center point of the suspended dome structure sits on the center support.

In the unloading construction stage, 81 temporary supports for positioning and supporting of members set up in the installation construction stage were removed. In line with the principle of "deformation coordination, unloading balance," the specific order of unloading was as follows. The first step was to remove 36 outer-loop supports symmetrically. The second

Figure 5.6 Installation construction of the prestressed cables.

Figure 5.7 Removing construction of the middle-loop supports.

Figure 5.8 All of the temporary supports had been removed.

step was to remove the central support of the central position of the stadium and to adjust the prestressed cable to the designed installation length. The third step was to remove 36 middle-loop supports symmetrically in a clockwise direction. The fourth step was to remove inner-loop supports synchronously and step by step. The photos in the construction process are shown in Figure 5.7 and Figure 5.8, respectively.

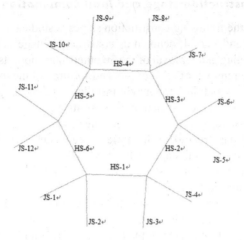

Figure 5.9 Schematic diagram for numbering the cables.

The construction of prestressed cable was carried out after the completion of unloading. After the start of tensioning, the first step was to synchronize the tensioning of js-1, js-5, and js-9 to 86kN, 86kN, and 86.1 kN. The second step was synchronous tensioning of js-2, js-6, and js-10 to 52.9 kN, 53kN, and 52.9 kN. The third step was synchronous tensioning of js-3, js-7, and js-11 to 116.2 kN, 116.3 kN, and 116.2 kN. The fourth and final step was to synchronize the tensioning of js-4, js-8, and js-12 to 105kN. The cable numbers are shown in Figure 5.9.

3   Stability analysis of time-varying structure during construction

Steel spatial hybrid structure stadium construction is a complex structural system with gradual process of construction. During construction the structure experienced from partial to whole, from simple to complex, from installation configuration to final configuration. The structure itself has undergone a series of changes as the construction progresses, and it belongs to the category of time-varying structure mechanics research, so its stability analysis is more complex (Simtses, 1990; Mateescu et al., 2000). The overall stability of the structure was analyzed by using the general finite element software MIDAS/Gen. According to the design data and construction organization design, the following steps were involved: defining the relevant properties of structural members, determining the boundary conditions and load conditions, dividing the construction stage, and establishing a structural model. The buckling analysis was performed using the "buckling analysis control" function in the software to obtain the buckling modes and critical buckling coefficients of the structure. Using the "displacement control method" in "nonlinear analysis control," the geometric nonlinear buckling analysis of the structure is performed to obtain the load-displacement curve of the structure under the corresponding load conditions, and the corresponding critical buckling coefficient. Then the stability of the structure construction process was checked to determine whether it satisfied the safety requirements.

3.1   Determine construction stage and load combination

The overall stability of the following construction stages is studied: (1) the closing of structural outer ring truss 3 and truss 4, denoted as construction stage 1; (2) the completion of prestressed cable tensioning and removal of the 36 outer-loop supports, denoted as construction stage 2; and (3) the removal of all supports and forming of the main structure, denoted as construction stage 3. In addition, the newly installed single-span truss is the most unfavorable substructure during the construction because its adjacent substructures are not yet installed, and the sidewise restraints are weak. It is necessary to use linear buckling analysis to check the stability of the structure at this time to see whether it meets the requirements. For this purpose, the construction stage of the first HJ3 + HJ4 installation is recorded as construction stage A; the construction stage of the first HJ1 + HJ2 installation is completed is recorded as construction stage B.

In order to simulate the construction load, the load combination in construction stages 1, 2, and 3 is dead load + live load + wind load, application of concentrated loads of 1kN in the vertical direction to the joints at the top of the first HJ3 + HJ4 and HJ1 + HJ2 in construction stages A and B.

3.2   Linear buckling analysis

It is difficult to accurately reflect the stable bearing capacity of the structure through linear buckling analysis. However, this method has a clear concept and is easy to calculate, which helps to understand the overall stability of the structure. It is of great significance to study the structural stability during construction by linear buckling analysis. Before conducting the nonlinear stability analysis of the whole structure, linear buckling analysis is necessary (Zhao et al., 2005; Jiang et al., 2013). Table 5.1 shows the critical buckling coefficient of each order of linear buckling obtained by finite element analysis in the five construction stages. The first-order buckling mode diagram corresponding to each stage is shown in Figure 5.10.

The results of linear buckling analysis show that the stability of structures in five dangerous construction stages all meet the safety requirements.

Table 5.1 The critical buckling coefficient of each order

Construction stage	Mode 1	Mode 2	Mode 3	Mode 4	Mode 5	Mode 6
1	63.49	64.36	69.96	71.23	73.81	75.81
2	16.51	17.37	19.53	23.60	25.71	26.53
3	30.31	30.51	30.88	31.11	31.41	31.52
A	241.3	490.8	1166.5	1419.2	–	–
B	393.4	653.2	953.4	1252.9	–	–

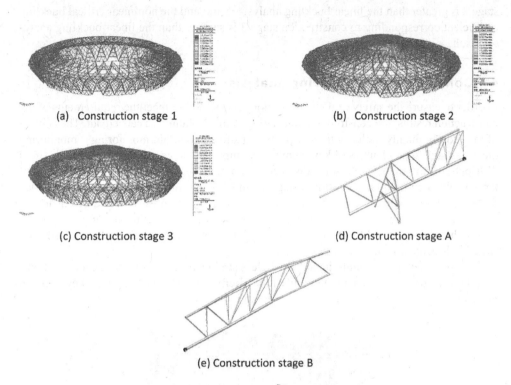

(a) Construction stage 1

(b) Construction stage 2

(c) Construction stage 3

(d) Construction stage A

(e) Construction stage B

Figure 5.10 The first-order buckling mode diagram corresponding to each stage.

Table 5.2 The nonlinear critical buckling coefficient of each construction stage

Construction stage	1	2	3
Critical buckling coefficients	68.76	18.83	27.56

3.3   Nonlinear buckling analysis

The complex large-span steel structure system usually presents obvious nonlinearity and is usually a defect-sensitive structure. Linear buckling analysis cannot examine the influence of initial defects on the structural stability, and it is necessary to perform a nonlinear full-process analysis on the overall structure (Zhao et al., 2005; Jiang et al., 2013). For the nonlinear buckling analysis of the steel spatial hybrid structure stadium, only the influence of geometric nonlinearity is considered. Table 5.2 shows the critical buckling coefficients of the overall structure, which are obtained from finite element analysis results, for construction stages 1, 2, and 3.

The results of nonlinear buckling analysis show that the overall stability of the structure meets the safety requirements during the three construction stages of the study. The nonlinear critical buckling coefficient corresponding to construction stage 1 and construction stage 2 is greater than the linear buckling analysis result; and the nonlinear critical buckling coefficient corresponding to construction stage 3 is smaller than the linear buckling analysis result.

4   Construction monitoring analysis

In order to ensure the safety of the construction process and to meet the requirements of the construction results, the construction monitoring of the stadium was carried out. The content of monitoring mainly includes three aspects: (1) stress and strain monitoring – monitoring the stress and strain changes of key members at important stages and analyzing their stress distribution and safety reserves, (2) displacement monitoring – monitoring the deflection of key members and joints at important stages, and (3) cable force monitoring of suspended dome structures – monitoring the real-time cable force of the prestressed cable tensioning process to check and control the safety and accuracy of the tensioning construction, and to ensure that the structural force of the suspended dome structure is reasonable and the construction is economical.

The equipment used mainly included wireless strain gauges (Figure 5.11), total stations, a cable force diagnosis system, and other tools. Reflectors are attached to the key joints to monitor the deflection of the point, as shown in Figure 5.12.

Figure 5.11 Wireless strain gauge.

Figure 5.12 Ball joint with reflector.

4.1   Monitoring point layout

According to the established construction plan, the entire process of construction in the stadium was simulated by finite element analysis. Through the analysis results, members with high stress level and large stress change range, and the joints with large space movement distance and complex deformation characteristics, can be found. Then the monitoring point layout plan can be determined based on the characteristics of the structural construction scheme and the operability of construction monitoring (Bruant et al., 2001; Reynier & Abou-Kandil, 1999). The construction of prestressed cable tensioning is the key, and difficult, point in the construction of the suspended dome structure. The forming cable force of any one cable has an important influence on the safety of the structure. Therefore, the 18 prestressed cables of this structure are all monitored.

The stress monitoring points consists of three parts, which are respectively recorded as zone 1, zone 2, and zone 3 (as shown in Figure 5.13, 5.14, and 5.15). Each zone is arranged

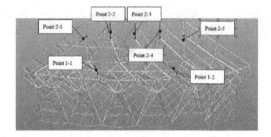

Figure 5.13 Arrangement plan for stress monitoring points of zone 1.

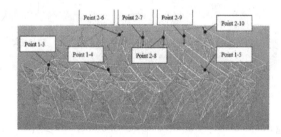

Figure 5.14 Arrangement plan for stress monitoring points of zone 2.

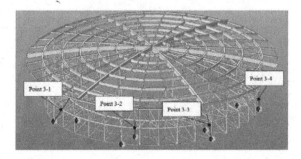

Figure 5.15 Arrangement plan for stress monitoring points of zone 3.

with 7, 8, and 4 monitoring points, a total of 19 points. It is used to monitor the stress values and changes of key members in the construction process. The displacement monitoring points are composed of six parts, respectively denoted as loop 1, loop 2, and loop 3 (as shown in Figure 5.16), loop 4 (as shown in Figure 5.17), and loop 5 (as shown in Figure 5.18), and dome central displacement points (as the 123 joints of the suspended dome structure), to monitor the spatial displacement of key members and joints during construction. Cable force monitoring is performed on all cables.

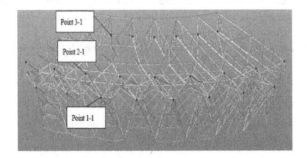

Figure 5.16 Arrangement plan for stress monitoring points of loops 1, 2, and 3.

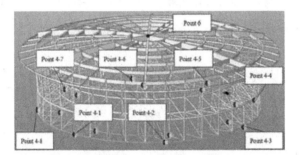

Figure 5.17 Monitoring points of loop 4 and the center of dome.

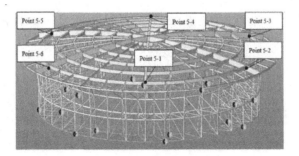

Figure 5.18 Arrangement plan for monitoring points of loop 5.

4.2   Analysis of monitoring results

By comparing the measured value with the theoretical value obtained by the finite element software analysis, the actual state of the hybrid space steel structure gymnasium construction can be studied.

1.   The measured results of the three stress monitoring areas show that the measured values of most of the members are in good agreement with the theoretical values. The values of the two components have consistent variation and similar values. Figure 5.19 shows the comparison between the measured values of the vertical web member in zone 2 and the theoretical calculations of the finite element theory. However, there are also some differences between the measured values of the members and the theoretical values (Figure 5.20). It is believed that this phenomenon occurs because either these bars are greatly affected by the construction, the uncertainties are many, or the stress level of the bars themselves is low. The monitoring results show that the stress level of the structure during the whole construction process meets the requirements of construction stability and stress control.
2.   The results of displacement monitoring show that the displacement of most monitoring points is consistent with the overall displacement of the structure, and the vertical displacement accounts for a large proportion of the total spatial displacement.
       However, the vertical displacement of the outermost 36 ball joints in the early part of the construction stage is negative; that is, the case where the monitoring point moves

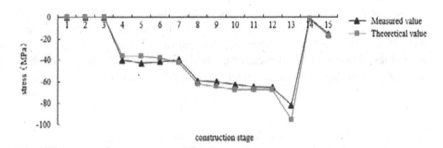

Figure 5.19 Vertical member stress comparison chart.

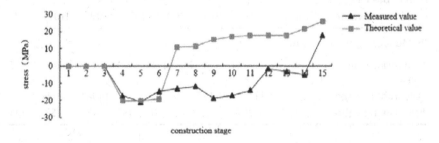

Figure 5.20 Large difference in stress.

downward. It is indicated that some of the monitoring points moved upwards in the early stage of construction and are separated from the outer-loop support. The main reason for this phenomenon is that the welding process has caused the shrinkage of the members. As the installation process of the members is continuously carried out, the landing part of the truss is subjected to the pulling force of the remaining bars and rotates around the support to the central axis of the stadium, thereby causing the measuring point to continuously rise.

At the same time, the vertical displacement of the joints at the top of the middle-loop supports do not show a consistent change with the total displacement of the space. Due to the assembly error and the influence of the welding process, the spatial movement laws of different points are different; however, the displacement values of the points are small. After the middle-loop supports are completely removed, the monitoring points of loop 3 move downwards as a whole, the moving range is large, and the vertical displacement accounts for a large proportion of the total displacement. In addition, for the connection joints of the inner-loop grid and the corresponding inner-loop supports, theoretically, these joints cannot be translated or rotated because they are connected by welding. But, the monitoring results show that the points have moved spatially, and the vertical downward movement plays a leading role. The reason is that the inner-loop supports bear a large load and sit directly on the ground. The ground soil has a certain compressibility, which causes the monitoring points of loop 4 to have a vertical downward displacement before the inner-loop supports are removed. The rigidity of the inner-loop supports are small, and the constraint on the monitoring points of loop 4 are limited, which causes the points of loop 4 to have displacement in the horizontal plane during the installation construction.

The monitoring results show that the displacement of each joint in the steel spatial hybrid structure stadium meets the stability and displacement control requirements.

3. After the construction of the suspended dome structure, the theoretical values, monitoring values, and errors of all prestressed cable forces are shown in the Table 5.3. The monitoring results of cable force meet the design requirements.

Table 5.3 The critical buckling coefficient of each construction stage

(a) Monitoring point 1–1 to 1–9

Monitoring point number	1–1	1–2	1–3	1–4	1–5	1–6	1–7	1–8	1–9
Allowable range					100kN ~ 110kN				
Monitoring value	109	101	104	100	105	109	110	104	109

(b) Monitoring point 1–10 to 2–6

Monitoring point number	1–10	1–11	1–12	2–1	2–2	2–3	2–4	2–5	2–6
Allowable range	100kN ~ 110kN				140kN ~ 154kN				
Monitoring value	108	107	100	100	142	150	153	143	143

5 Conclusion

1. The structure of the stadium considered in this chapter is a steel spatial hybrid structure system with a large span. It is essential to analyze and monitor the overall stability of the construction process. The results of such analysis and monitoring can provide a theoretical basis for the safety study of the structure.

2. The construction method in installation construction stage is the separate hoisting method. After the installation of the members, the temporary supports are removed according to the principle of "deformation coordination, unloading balance." The tension of the prestressed cable is carried out after the unloading is completed, and the radial cable is selected as the active cable.

3. The results of linear and nonlinear buckling analysis show that the overall structural stability of each construction stage meets the safety requirements. The critical buckling coefficient obtained in construction stages 1, 2, and 3 of the nonlinear buckling analysis is close to the linear buckling analysis.

4. The monitoring results show that the displacement of each joint meets the stability and displacement control requirements during the construction process of the steel spatial hybrid structure stadium. However, due to the influence of various factors during the construction of complex steel structures, there are some differences between the stress state and the theoretical value in some areas during construction.

References

Borri, C. & Spinelli, P. (1988) Buckling and post-buckling behavior of single reticulated shells effected by random imperfection. *Comput & Struct*, 30(4), 937–943.

Bruant, I., Coffignal, G., Lene, F. & Verge, M. (2001) A methodology for determination of piezoelectric actuator and sensor location on beam structures. *Journal of Sound and Vibration*, 243(5).

Jian Yuan, Huan-ting Zhou & Cang-ru Jiang (2009) Stability analysis of long-span arch truss structure with circular steel tube. *Building Science*, 25(1), 86–90.

Maoyuan Gao, Jian Li, & Yongfeng Luo (2012) Stability analysis of peculiar steel structure. *Structural Engineers*, 28(6), 14–18.

Mateescu, D., Gioncu, V. & Dubina, D. (2000) Timisoara steel structures stability research school: Relevant contributions. *Journal of Constructional Steel Research*, 55(1).

Mingyong Jing, Yuanqing Wang, Yong Zhang, Yongjiu Shi & Zhenyi Bian (2008) Design and analysis of large span spatial steel tubular gymnasium structure. *Building Structure*, 38(2), 16–18, 96.

Reynier, M. & Abou-Kandil, H. (1999) Sensors location for updating problems. *Mechanical Systems and Signal Processing*, 13(2).

Simtses, G.J. (1990) *Dynamic stability of suddenly loaded structures*. Springer-Verlag, New York.

Yang Zhao, Xianchuan Chen & Shilin Dong (2005) Strength and stability analysis of circular steel arches in a long-span ellipsoidal shell structure. *China Civil Engineering Journal*, 38(5), 15–23.

Zhengrong Jiang, Shitong Wang, Kairong Shi, Zhihan Peng, Xiaonan Gao & Peng Kong (2013) Nonlinear buckling analysis of long-span elliptic paraboloid suspended dome structure for Houjie Gymnasium. *China Civil Engineering Journal*, 46 (9), 21–28.

Chapter 6

Structural health monitoring of some space frames in China

Jinzhi Wu, J. Hu, Y. G. Zhang, M. L. Liu, and X. W. Song

1   Introduction

With the rapid development of the economy and the progress of the structural design and construction technology, more and more large projects have been built in recent years in China. The demand for assessment and evaluation of the existing and newly built structures for hazard mitigation have significantly expanded the research efforts in the field of SHM. In recent years, SHM has been applied to more and more bridges and large buildings, in which environmental and structural parameters are measured to assess the state of the structure. For such large structures, the static parameters of the structures are difficult or impossible to measure, while the modal parameters (modal frequency, damping ratio, and mode shape) of the structure are quite easy to obtain. Operational modal analysis (OMA) is the most practical and efficient method to obtain the modal parameters, which can be identified by different OMA methods based on the vibration information measured via accelerometers placed on the structures (Brinker and Ventura, 2015). This chapter extensively reviews the OMA methods, including frequency domain methods and time domain methods (Reynders, 2012). Different methods are applied to practical projects, and the results have been compared with each other to show the merits and drawbacks of different methods. The SHM results helped to assess the condition of the measured structures, and based on that assessment some recommendations have been given. It is hoped that these recommendations will help the application of SHM to large spatial structures.

2   Operational modal analysis methods

Experimental modal analysis (EMA) is the fundamental method to obtain the modal parameters of structure in the laboratory, which rely on the input excitation and output vibration information of the structure system (Brinker and Ventura, 2015). But for the practical large projects, it is difficult or impossible to excite the structure, and the practical method is to analyze the vibration information of the structure under ambient excitation, where the structure is in the working or operational state. This priority leads to the development of different OMA techniques. According to the analytical domain, OMA can be divided into three categories: frequency domain methods, time domain methods, and time-frequency domain methods.

2.1   Frequency domain identification

Quite a few different methods have been developed over the years, especially during the 1970s and the 1980s. The frequency domain method is based on Fast Fourier Transform (FFT) (Cooley and Tukey, 1965).

The classical frequency domain approach, also known as the "Peak Picking" approach (PP), has been used for modal identification for decades, in which the natural frequencies were evaluated directly from the peaks of the frequency response functions (FRFs) and the damping ratios were calculated by the half-power points method (Bendat and Piersol, 1993). Zhang (Zhang et al., 2013) proposed an improved power spectrum peak method applied in the modal identification of spatial lattice structures. It combines the average normalized power spectrum measured at all the test points with an auxiliary normalized power spectrum calculated according to the characteristics of theoretical mode shapes, and then the union set is defined as the final identification result. The advantage of the approach is its simplicity and high speed, but the problem is that it does not work well with closely spaced modes, which is a common characteristic of spatial structures.

The frequency domain decomposition technique (FDD) introduces the concept of singular value decomposition (SVD) on the basis of the PP method. This not only retains the advantages of the PP method for rapid identification, but also gives it the ability to resist noise and analyze cases with closely spaced modes (Brincker et al., 2000). The SVD of the spectral density (SD) matrix estimated from the random response can obtain a corresponding set of power spectra of the single-degree-of-freedom system. The frequency corresponding to the peak of each power spectrum curve is a certain frequency of the system. One of the advantages of the FDD is that it separates the noise from the physical meaning by SVD plots that clearly show what is noise and what is structural-related information. The damping ratio recognized by the FDD method is not quite accurate. Based on FDD, EFDD (Brincker et al., 2001) and FSDD (Wang and Zhang, 2006) were developed with some modification to FDD. But the damping ratio is still difficult for them to identify accurately.

2.2 Time domain identification

Time domain identification can be divided into two kinds according to the different calculation processes: one is to recognize directly by the structural response; the other is to preprocess the signal before recognition to obtain the correlation function or free decay response of the signal (Xu et al., 2002). Random Decrement Technique (RDT) (Cole, 1973) and Natural Excitation Technique (NExT) (James et al., 1995) are two kinds of preprocessing methods. One of the earliest time domain identifications is to identify the modal parameters by fitting the Auto-Regressive and Moving Average (ARMA) models to the measured signal (Akaike, 1969). The Ibrahim time domain (ITD) method introduced by Ibrahim (Ibrahim and Mikulcik, 1977) in the late 1970s is one of the first techniques developed for the modal identification of multiple output systems. The Least-squares Complex Exponential Method (LSCE) uses the relationship between the single impulse response function (IRF) and the residue and the pole to find the modal parameters of the structure (Brown et al., 1979). The Eigensystem Realization Algorithm (ERA) takes the IRF as the basic model and obtains the minimum implementation of the system through SVD (Juang and Pappa, 1985). Two main types of Stochastic Subspace Identification (SSI) formulations are the correlation-driven SSI and the data-driven SSI (Peeters et al., 1995). The theoretical foundation of the SSI is the time-domain state-space equation. SSI is widely used for its solid theoretical basis and identification results for the modal parameters.

2.3 Time-frequency domain identification

The Fourier transform cannot reflect the frequency change over time in the time-varying signal. Time-frequency analysis, such as Wavelet Transform (Xu *et al.*, 2003) and Hilbert-Huang Transform (Huang *et al.*, 1998), can obtain spectrum changes in different time periods or time-domain amplitude changes in different frequency bands, which can overcome the shortcomings of traditional analysis methods in time expression. The time-frequency domain identification can effectively deal with nonstationary signals, but the recognition efficiency is low. It is necessary to rely on the Fourier transform to determine the range of the modal frequencies of each order.

3 SHM application to spatial structures

In this section, the SHM is applied to several spatial structures for the assessment for their state. The types and numbers of sensors, data analysis methods, and the results are introduced. Structural vibration is measured under ambient excitation, which mainly comes from the wind, the ground microtremor, and so on.

3.1 Shijiazhuang Yutong International Stadium

The roof of the stadium is a square pyramid space frame with welded hollow spheres. The roof is 120 meters long, and the cantilever is 47 meters. The project is in Hebei Province, as shown in Figure 6.1.

According to the numerical analysis results and the practical engineering characteristics, 16 piezoelectric accelerometers were placed at the lower chord nodes, including 12 vertical sensors, 2 lateral sensors, and 2 longitudinal sensors. The layout of the sensors is shown in Figure 6.2. The measured results are analyzed using the OMA methods outlined earlier and

Figure 6.1 Shijiazhuang Yutong International Stadium.

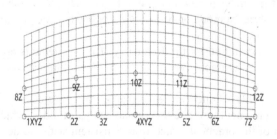

Figure 6.2 Layout of sensors.

Table 6.1 Frequencies and damping ratios of Shijiazhuang Yutong International Stadium

Method	Mode 1		Mode 2		Mode 3		Mode 4		Mode 5		Mode 6	
Parameters	f_n	ξ_n	f_n	ξ_n	f_n	ξ_n	f_n	ξ_n	f_n	ξ_n	f_n	ξ_n
	/Hz	/%	/Hz	/%	/Hz	/%	/Hz	/%	/Hz	/%	/Hz	/%
FEA	3.0	–	3.25	–	3.66	–	3.74	–	4.06	–	4.75	–
PP.	2.0	–	2.25	–	3.06	–	3.50	–	3.94	–	4.38	–
FDD	2.0	–	2.25	–	3.06	–	3.50	–	3.94	–	4.38	–
ARMA	1.8	12.7	2.24	1.12	3.08	5.05	3.46	1.90	4.08	7.96	4.29	1.80
ITD	1.9	6.3	2.25	1.47	3.03	4.90	3.50	2.75	4.06	6.01	4.35	1.62
ERA	2.0	4.1	2.33	5.25	3.02	3.22	3.46	3.30	3.85	4.67	4.32	1.40
SSI-DATA	1.9	7.3	2.21	3.18	3.00	4.47	3.48	5.24	3.94	7.91	4.35	1.16
SSI-COV	1.9	4.0	2.26	1.09	3.01	3.45	3.47	3.58	4.05	5.66	4.33	1.32

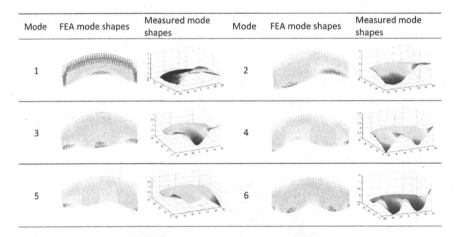

Figure 6.3 The first six mode shapes of the Shijiazhuang Yutong International Stadium by SSI-COV method.

compared with the finite element analysis (FEA) results by SAP2000 of the structure (Xu, 2014). The results of frequencies and damping ratios are shown in Table 6.1. Due to space limitations, only the vibration mode obtained by the SSI-COV method appears in this chapter, as shown in Figure 6.3.

The damping ratios identified by the frequency domain methods are not good enough, and the calculation results are not given here. A dash ("–") indicates that the mode is not identified clearly.

It can be seen from Table 6.1 and Figure 6.3 that the first six measured mode shapes of the stadium are in good agreement with the theoretical analysis modes, which indicates that the stiffness distribution of the practical structure is consistent with the design. The frequency identification results of each method are relatively close, but the damping ratios are quite different. Among them, the frequency domain method has the fastest speed and works well for the modal frequencies and mode shapes. In this project, the measured frequencies are lower than those of FEA. It is because the stadium has been built for many years, and because the original drawings and the design documents are incomplete, which leads to FEA errors. The measured data will be used for the model updating of the existing structure, and the monitoring results will be the basis for future structural monitoring and state assessment.

3.2 A hangar in Xi'an

The hangar is a three-layer square pyramid space frame with welded hollow spheres. It is 138.6 meters long, 63.7 meters wide, and 30 meters high, which is in Shanxi Province. The entrance edge of the roof is reinforced by a four-layer truss (Figure 6.4). The layout of the accelerometers is shown in Figure 6.5, including 14 vertical sensors, 1 lateral sensor, and 1 longitudinal sensor. The measured results are analyzed using different methods and compared with the FEA results of the structures (Hu, 2017). The results of frequencies and damping ratios are shown in Table 6.2. The mode shapes obtained by the SSI-COV method are given in Figure 6.6.

It can be seen that the ITD method and the ERA method cannot identify the fourth-order modal parameters of the structure. The third-order damping ratio by ARMA method is not good because it uses only the data of one measuring point for analysis. The SSI-COV method is more accurate compared to other methods. As the modal order increases, the discretization of the modal parameters also increases. The first several mode shapes can be quite clearly identified and similar with the FEA results, but there is a significant difference between the frequencies by different methods.

Figure 6.4 A hangar in Xi'an.

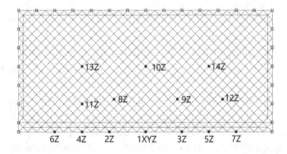

Figure 6.5 Layout of sensors.

Table 6.2 Frequencies and damping ratios of the hangar in Xi'an

Method	Mode 1		Mode 2		Mode 3		Mode 4	
Parameters	f_n	ξ_n	f_n	ξ_n	f_n	ξ_n	f_n	ξ_n
	/Hz	/%	/Hz	/%	/Hz	/%	/Hz	/%
FEA	1.72	–	1.84	–	2.21	–	2.83	–
PP	1.69	–	3.00	–	3.50	–	3.75	–
FDD	1.69	–	2.81	–	3.50	–	3.75	–
ARMA	1.72	1.78	2.81	1.37	3.23	4.83	3.75	16.6
ITD	1.72	0.24	2.85	1.72	3.62	3.10	–	–
ERA	1.71	0.10	2.85	3.65	3.28	7.98	–	–
SSI-DATA	1.72	0.26	2.84	3.87	3.66	5.81	3.96	5.96
SSI-COV	1.72	0.50	2.82	1.59	3.49	3.48	3.81	6.71

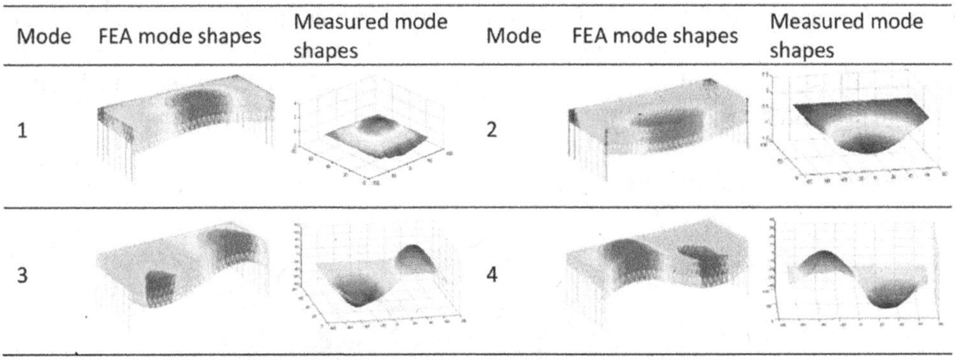

Mode	FEA mode shapes	Measured mode shapes	Mode	FEA mode shapes	Measured mode shapes
1			2		
3			4		

Figure 6.6 First four mode shapes of the hangar in Xi'an.

3.3   A space frame in Xuzhou

The square pyramid space frame with welded hollow spheres is 96.0 meters long, 87.6 meters wide, and 11.1 meters high, and is in Jiangsu Province (Figure 6.7). There are 14 vertical sensors, 1 lateral sensor, and 1 longitudinal sensor used, as shown in Figure 6.8. The measured data are analyzed using the above method and compared with the FEA results of the structures (Hu, 2017). The results of frequencies and damping ratios are shown in Table 6.3. The mode shapes obtained by the SSI-COV method are given in Figure 6.9.

Among the above identification results, it can be seen that ambient vibration failed to excite the first-order mode shape of the frame, while the ARMA method failed to identify the sixth-order mode parameters. And the fourth- and fifth-order damping ratios identified by the ITD method are quite different from other methods. It should be noted that the measured

Figure 6.7 A space frame in Xuzhou.

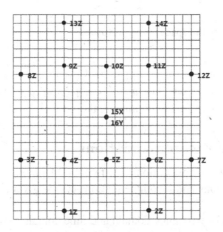

Figure 6.8 Layout of sensors.

Table 6.3 Frequencies and damping ratios of a space frame in Xuzhou

Method	Mode 1		Mode 2		Mode 3		Mode 4		Mode 5		Mode 6		Mode 7	
Parameters	f_n	ξ_n	f_n	ξ_n	f_n	ξ_n	f_n	ξ_n	f_n	ξ_n	f_n	ξ_n	f_n	ξ_n
	Hz	%	Hz	%	Hz	/%	Hz	%	Hz	%	Hz	%	Hz	%
FEA	2.5	–	2.6	–	3.1	–	5.3	–	6.4	–	6.9	–	9.9	–
PP	–	–	1.9	–	2.1	–	2.8	–	3.0	–	3.5	–	4.1	–
FDD	–	–	1.9	–	2.1	–	2.8	–	3.0	–	3.5	–	4.1	–
ARMA	–	–	1.9	2.9	2.0	6.24	2.9	6.1	3.0	5.5	–	–	4.0	3.8
ITD	–	–	1.8	6.1	2.0	2.68	2.9	6.7	3.1	6.7	3.5	3.6	4.0	2.8
ERA	–	–	1.9	6.4	2.1	3.63	2.8	3.7	3.1	1.7	3.4	4.4	4.1	1.2
SSI-DATA	–	–	1.9	4.9	2.1	2.94	2.8	3.4	3.1	2.6	3.3	3.5	4.1	3.5
SSI-COV	–	–	1.8	4.3	2.1	2.41	2.8	2.9	3.1	2.7	3.4	4.4	4.1	2.9

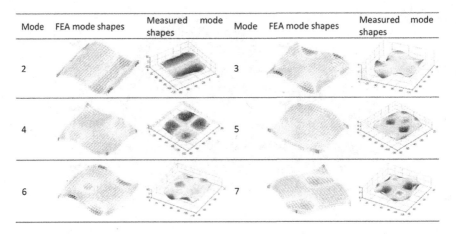

Mode	FEA mode shapes	Measured mode shapes	Mode	FEA mode shapes	Measured mode shapes
2			3		
4			5		
6			7		

Figure 6.9 2–7 mode shapes of the space frame in Xuzhou.

frequencies are quite lower than those of FEA. The structure will be monitored again in the near future in order to confirm whether stiffness reduction of the structure is taking place. The model updating is also needed to obtain a real model of the structure.

4   Conclusion and recommendations

Based on the introduction and application of different OMA methods presented in this chapter, the following conclusions and recommendations can be given:

1. The theoretical bases of frequency domain and time domain OMA methods are different, which leads to different results. The application of these methods indicates that the frequency domain methods work well enough for the modal frequency and the mode shape, and that the SSI method is more accurate in frequency, damping, and mode shape estimate.
2. Even for large structures, fewer than 30 sensors seem enough to measure the first few mode shapes.
3. The standard deviations obtained for the damping ratios are much higher than those obtained for the modal frequencies, especially for frequency domain methods. Time domain methods are recommended for damping ratios.
4. The modal parameters of a structure may be affected significantly by temperature and other environmental factors, which should be considered during SHM.
5. The measured modal parameters have significant difference with the FEA values for some projects. This should not be regarded as a failure of the SHM. This, in fact, is an important step of SHM. The measured modal parameters provide the base information of the existing projects, which is essential for further monitoring, model updating, damage diagnosis, and state assessment of the structure in the future.

Acknowledgements

This research was funded by the National Natural Science Foundation of China (51278009).

References

Akaike, H. (1969) Power spectrum estimation through autogressive model fitting. *Annals of the Institute of Statistical Mathematics*, 21, 243–247. Available from: https://link.springer.com/article/10.1007%2FBF02532269?LI=true

Bendat, J.S. & Piersol, A.G. (1993) *Engineering Application of Correlation and Spectral Analysis*, 2nd ed. John Wiley & Sons, Inc., New York, USA.

Brinker, R. & Ventura, C. (2015) *Introduction to Operational Modal Analysis*. John Wiley & Sons, Ltd., West Sussex, UK.

Brincker, R., Zhang, L. & Andersen, P. (2000) Modal identification from ambient responses using frequency domain decomposition. *Proc. 18th Int. Modal Analysis Conf. (IMAC), SanAntonio, Tex, USA*.

Brincker, R., Ventura, C. & Andersen, P. (2001) Damping estimation by frequency domain decomposition. *Proceedings of 19th International Modal Analysis Conference, Society for Experimental Mechanics, Kissimmee, USA*.

Brown, D., Allemang, R., Zimmerman, R., *et al.* (1979) *Parameter Estimation Techniques for Modal Analysis*. SAE Technical Paper No. 790221. Available from: https://doi.org/10.4271/790221

Cole, H.A. (1973) *On Line Failure Detection and Damping Measurement of Aerospace Structure by Random Decrement Signature*. Nielson Engineering and Research, Inc. Report number NASA CR-2205, Mountain View, CA, USA.

Cooley, J.W. & Tukey, J.W. (1965) An algorithm for the machine calculation of complex Fourier series. *Mathematics of Computation*, 19(90), 297–301.

Hu, J. (2017) *Comparison of Structure Operational Modal Analysis Methods and Software Implementation*. Dissertation of Beijing University of Technology, China.

Huang, N.E., Shen, Z., Long, S.R., et al. (1998) The empirical mode decomposition and the Hilbert spectrum for nonlinear and non-stationary time series analysis. *Proceeding of the Royal Society of London: Series A: Mathematical, Physical and Engineering Sciences*, 454(1971, March), 903–995.

Ibrahim, S.R. & Mikulcik, E.C. (1977) A method for the direct identification of vibration parameters from the free response. *Shock and Vibration Bulletin*, 47, 183–198.

James, G.H., Carne, T.G. & Lauffer, J.P. (1995) The natural excitation technique (NExT) for modal parameter extraction from operating structures. *The International Journal of Analytical and Experimental Modal Analysis*, 10(4), 260–277.

Juang, J.N. & Pappa, R.S. (1985) An eigensystem realization algorithm for modal parameter identification and modal reduction. *Journal of Guidance, Control, Dynamic*, 8, 620–627.

Peeters, B., De Roeck, G., et al. (1995) Stochastic subspace techniques applied to parameter identification of civil engineering structures. *Proceeding of New Advance in Modal Synthesis of large Structures: Nonlinear, Damped and Nondeterministic Cases, September, Lyon, France*, pp. 151–162.

Reynders, E. (2012) System identification method for (Operational) modal analysis: Review and comparison. *Archives of Computational Methods in Engineering*, 19, 51–124.

Wang, T. & Zhang, L.M. (2006) Frequency and spatial domain decomposition for operational modal analysis and its application. *Acta Aeronautica et Astronautica Sinica*, 27(1), 62–66.

Xu, Q.Z. (2014) *Study on Modal Identification and Sensor Placement for Spatial Grid Structure*. Dissertation of Beijing University of Technology, Beijing, China.

Xu, X.Z., Hua, H.X. & Chen, Z.N. (2002) An overview of modal parameter identification methods based on ambient excitation. *Journal of Vibration and Shock*, 21(3), 2–5.

Xu, X.Z., Zhang, Z.Y., Huan, H.X., et al. (2003) Time-varying modal parameter identification with time-frequency analysis methods. *Journal of Shanghai Jiao Tong University*, 16(3), 58–362.

Zhang, Y.G., Liu, C.W., Wu, J.Z., et al. (2013) Improved power spectrum peak method applied in the modal identification of spatial lattice structures. *Journal of Vibration and Shock*, 32(9), 10–16.

Chapter 7

Collapse analysis of space-frame gymnasiums damaged in 2016 Kumamoto earthquake

Shogo Inanaga, Toru Takeuchi, Yuki Terazawa, and Ryota Matsui

1 Introduction

On April 14 and 16, 2016, two large earthquakes exceeding a peak ground acceleration (PGA) of 10 m/s^2 and 15 m/s^2, respectively, hit Kumamoto city, Kyushu, Japan in succession. These earthquakes resulted in the collapse of large numbers of residential buildings and caused severe damage to several school gymnasiums. Figure 7.1 shows the two gymnasiums with space frame roofs, which suffered damage due to buckling, fracture, and falling of some members. As shown in Figures 7.2a–c, A-gymnasium (hereafter referred to as Gym. A) lost most of its transverse chord members because of buckling and fracture. The entire roof started to sink and longitudinal members were fractured at connection bolts in tension and fell onto the floor. As shown in Figures 7.3a–c, B-gymnasium (hereafter referred to as Gym. B) suffered damage due to buckling and falling of some members near the front RC wall, which is attributed to the out-of-plane response of the front walls as reported in past earthquakes (ISCTJ, 2015). However, the process of the falling of the members is not clear, and their collapse mechanisms need to be investigated.

2 Observed damages

Figure 7.4 shows the schematic of the building of Gym. A, and Figure 7.5 shows that of Gym. B. Both have double-layered, cylindrical steel space frame roofs, supported by two stories of RC frames. The perimeter of the RC frames of the second floor is approximately 8 to 10 m high, and they are cantilever frames with low stiffness/strength in the out-of-plane directions. The steel roof is connected to these cantilevered RC frames with anchor bolts and steel bearings. Bearings are pin-supported at the four corners, and the other bearings have slotted holes, which allow sliding along out-of-plane directions of RC frames.

The damages suffered by Gym. A after the earthquakes are summarized in Figures 7.6 and 7.7, and Figures 7.8 and 7.9 show the same for Gym. B. In Gym. A, the damages are mainly observed along the Y1-Y2 and Y5-Y6 zones. All the compressive upper chord members along the transverse directions connected to the bearings buckled, and half of them were fractured at local buckling positions near the centre because of low-cycle fatigue failure. Diagonal members next to the buckled chords had also buckled. Several lower chords along the longitudinal directions had fractured at the connection bolts, and five of them fell onto the floor. Bending cracks were also observed at the base of the cantilevered RC columns supporting the roof, which indicates that the out-of-plane response of these cantilevered RC frames was significant.

(a) Gymnasium A (Gym. A)   (b) Gymnasium B (Gym. B)

Figure 7.1 Damaged gymnasiums by the 2016 Kumamoto earthquake.

 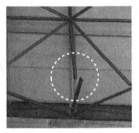 

(a) Buckling of the upper chord   (b) Fracture of the upper chord (c) Fell of the longitudinal members

Figure 7.2 Damages observed in Gym. A.

(a) Bolt fracture of lower chord (b) Buckling of diagonal chord (c) Buckling of lower chord

Figure 7.3 Damages observed in Gym. B.

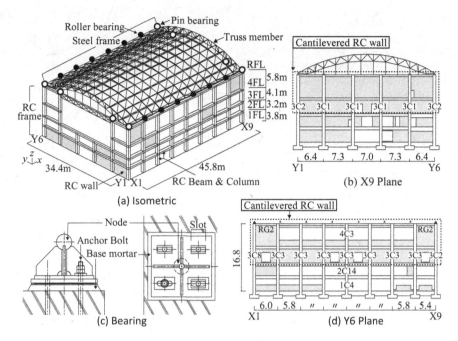

Figure 7.4 Building schematic of Gym. A.

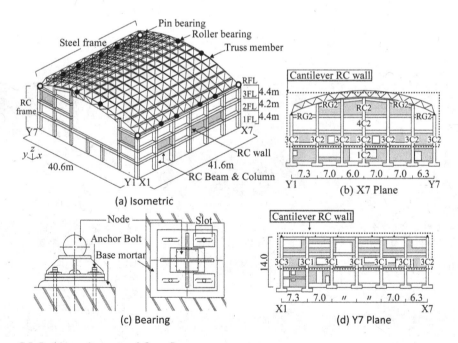

Figure 7.5 Building schematic of Gym. B.

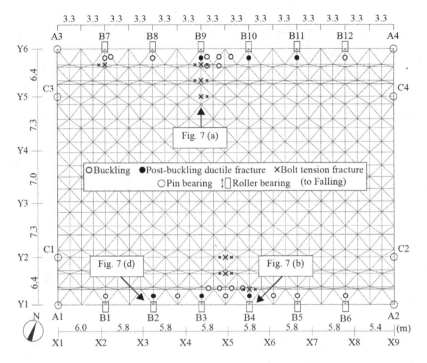

Figure 7.6 Roof grids of Gym. A.

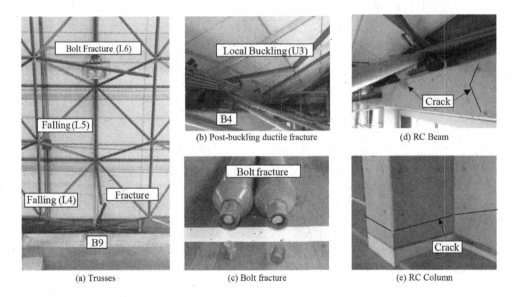

Figure 7.7 Observed damages in Gym. A.

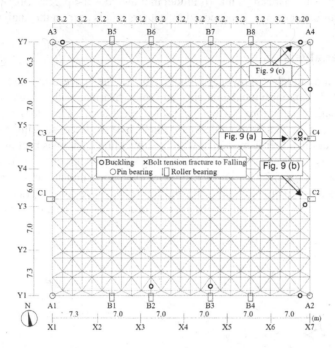

Figure 7.8 Roof grids of Gym. B.

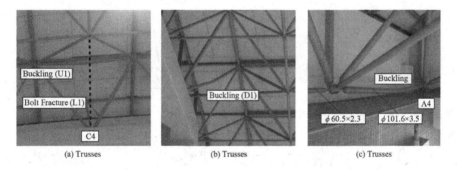

(a) Trusses   (b) Trusses   (c) Trusses

Figure 7.9 Observed damages in Gym. B.

In Gym. B, damages are mainly observed along the X7 side. Four members connected to the bearings buckled and one member fell on to the floor. The buckling of several chord members connected to bearings B2 and B3 along the transverse directions could also be observed.

The entire roof of Gym. A sank, and there was a risk of collapse. Some of the members fell to the floor and could have caused serious harm to people on the floor. Therefore, the process mechanism of such collapses needs to be researched. In the following, a detailed post-buckling and dynamic post-fracture numerical analyses are carried out to investigate the above collapse mechanism. A macro-model that includes the precise post-buckling hysteresis and fracture mechanism is proposed by authors and is used in the analysis.

3 Model analysis

The outline of the model is shown in Figure 7.10. All the space-frame members in each roof are modelled using individual elements, and roller bearings with slotted holes are modelled as expressing reaction forces when the displacement exceeds the slot length allowance. Hysteresis of RC members is modelled as bilinear hysteresis considering the after-crack stiffness. Each roof member is modelled with post-buckling hysteresis proposed by Shibata and Wakabayashi with the modification of buckling strength degradations (Taniguchi *et al.*, 1994), as shown in Figure 7.11a. Further, fracture evaluation under cyclic axial forces was carried out in each member using the strain amplitude amplification factors proposed earlier (Takeuchi & Matsui, 2015), and the stiffness and strength of the member is revised as zero in the analysis after the instance of fracture (Figure 7.11(b)).

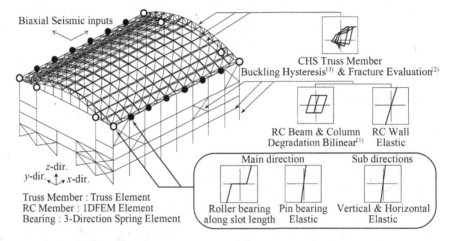

Figure 7.10 Numerical simulation model schematic (ex. Gym. A).

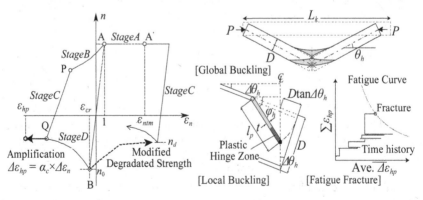

(a) Shibata-Wakabayashi buckling hysteresis   (b) Fracture evaluation model

Figure 7.11 Schematic image of post-buckling hysteresis and fracture evaluation method.

Seismic waves recorded at a point 0.5 km away from Gym. A and 2.4 km away from Gym. B were used for the analysis (Figure 7.12), and their angles were modified along the x-axis and y-axis of each gymnasium. There were two large earthquakes with PGAs of 10 m/s² on April 14 and 15 m/s² on April 16, which are called the pre-shock and main shock, respectively. In the analysis, both the records were applied continuously with a 10 s interval, as shown in Figure 7.12.

Figure 7.13 and Figure 7.14 show predominant modal shapes. In the x-direction of Gym. A, the coupled mode of the cantilevered RC frame and the roof excels in the pin support state, and the independent vibration mode of the cantilevered RC wall reported in the y-direction excels in the roller support state. In Gym. B, the independent vibration mode of RC frame

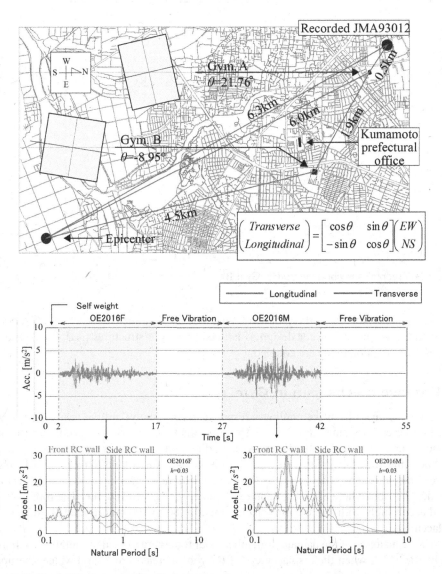

Figure 7.12 Used seismic record for damage evaluation.

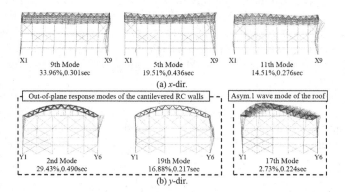

Figure 7.13 Predominant vibration modes (Gym. A).

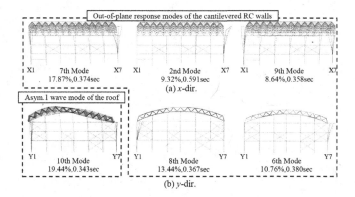

Figure 7.14 Predominant vibration modes (Gym. B).

is also dominant, but only in the y-direction, the roof anti-symmetric 1 wave mode reported by the AIJ recommendation for design of latticed shell roof structures (AIJ, 2016) and the independent vibration mode are coupled.

4   Damage mechanism in Gym. A

The results obtained from the nonlinear response history analysis of Gym. A, are summarized in Figures 7.15 and 7.16. Figure 7.15 shows the time history of the roof and RC wall displacements at bearings in the y- (transverse) direction along the Y1 and Y6 sides. In the pre-shock, the wall movement is generally within the slotted hole allowance (±44 mm), and roof displacement is also small. However, in the main shock, the wall displacement exceeded the slotted hole allowance and reached approximately 160 mm. The roof was displaced accordingly with a maximum displacement of 80 to 100 mm, forced by the bearing displacement.

The time histories of the damaged-member axial forces are plotted in Figure 7.16. Immediately after 30 s, when the displacement of the roof increased to a large value, the upper

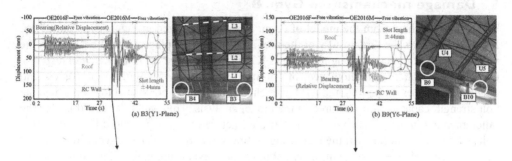

Figure 7.15 Time history of the bearing displacements (Gym. A).

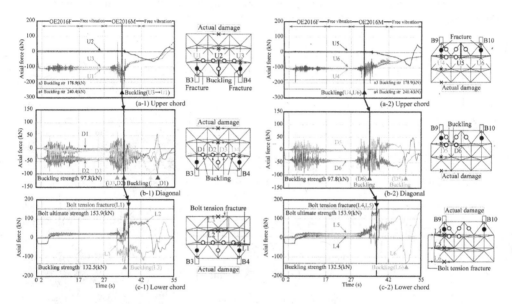

Figure 7.16 Time history of the axial member forces (Gym. A).

chord members (U1 and U3 in (a-1), and U4 and U6 in (a-2)) reached their buckling limit and thus lost their bearing capacity. After a few more seconds after the 30-s period, the diagonal members (D1–D4 in (b-1), and D5 and D6 in (b-2)) also reached their buckling limit and lost their bearing capacity. Immediately after these events, tensile forces in longitudinal lower chord members (L1 in (c-1) and L4 in (c-2)) reached the strength limit of the connection bolts and fractured. This process indicates that the load path changed from the transverse direction to the longitudinal direction as a result of the buckling of upper chord and diagonal members. Then the bolts of the longitudinal lower chord members fractured because of the introduction of the tensile forces, which was not considered in the design stage. It is assumed that this process led to the falling of members onto the ground. During the free vibrations after the main shock, upper chords U3 in (a-1) and U6 in (b-1) also reached their fracture limit, which was also observed in the actual damage.

5   Damage mechanism in Gym. B

The results obtained from the model of Gym. B are summarized in Figures 7.17 and 7.18. Figure 7.17 shows the time history of the roof and RC wall displacements at bearings in the x- (longitudinal) direction along the X1 and X7 sides. In Figure 7.17, the RC wall response exceeded the slotted hole allowance (of ±50 mm) in the early stage of the pre-shock, and tensile axial forces in the longitudinal lower chord member L1 connected to C4 reached the bolt strength capacity and fractured in the model analysis. This particular chord is also the fallen member in actual damage. Soon after the fracture of L1, upper chord U1 in the same unit as L1 buckled because of the increasing compressive forces. The transverse member L2 also buckled in this early stage of the pre-shock. These events are mainly caused by the out-of-plane response of the heavy RC frame at the X7 side in the x- (longitudinal) direction. In the main shock, the asymmetric-mode response of the roof in the y- (transverse) direction was more severe than the RC wall response, approximately 30 s; therefore, diagonal member D1 connected to bearing C2 also buckled.

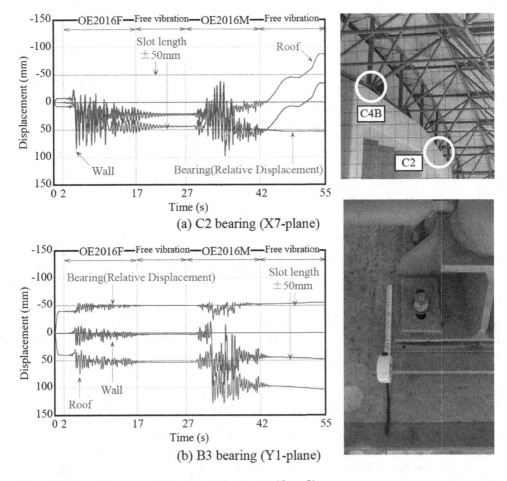

Figure 7.17 Time history of the bearing displacements (Gym. B).

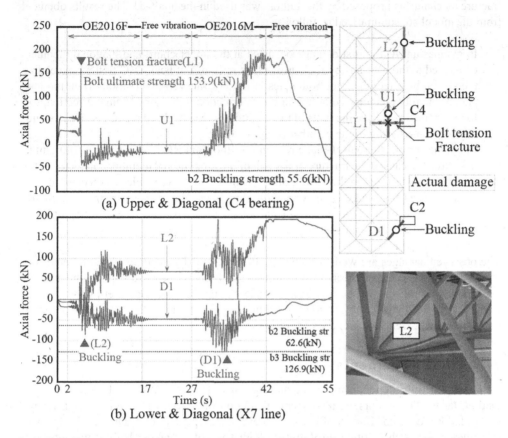

Figure 7.18 Time history of axial member forces related to the x-direction response (Gym. B).

The other buckled members along lines Y1 and Y7 also reached their buckling limit in the model analysis during both the pre-shock and main-shock period. In Gym. B, it can be said that the damages to the members close to side X7 were caused mainly by the out-of-plane response of the RC wall during the pre-shock period, and that damage to other members along sides Y1 and Y7 was caused mainly by the excitation response of the steel roof itself during the pre-shock and the main-shock period. The responses of the raised roofs are known to include asymmetrical vertical excitations, as indicated in the AIJ recommendation for latticed roof structures (AIJ, 2017). This design guidance also describes the out-of-plane response of supporting RC walls and is applicable to screening such types of damages in the future.

6 Summary

Detailed post-buckling and dynamic post-fracture numerical analyses were carried out to investigate the collapse mechanism of damaged gymnasiums in the 2016 Kumamoto earthquake. A macro-member model that includes the precise post-buckling hysteresis and fracture mechanisms proposed by the authors was used in the analysis. The results obtained from the model are summarized as follows:

1. In gymnasium A, the first damage occurred at the upper chord and diagonal members connected to the bearings by the out-of-plane response of heavy RC frames, exceeding the slotted hole length along both sides in the y- (transverse) direction. Then the load path of the roof changed from the transverse direction to the longitudinal direction, followed by the fracture of the bolts of the longitudinal lower chord members. This eventually led to the falling of the members onto the floor.
2. In gymnasium B, the out-of-plane response of the front RC frame in the x- (longitudinal) direction caused the fracture of the bolt of the longitudinal member connected to the bearing. This resulted in the falling of the member. The forces released because of this falling caused the buckling of surrounding members. Independently, the asymmetric response of the excited roof in the y- (transverse) direction caused the buckling of lower chord members along both sides.

The observed damages are well explained by the post-buckling and post-fracture analyses, which explain the collapse mechanisms of these gymnasiums and also validate the analysis models. More research will be required for generalizing the collapse pattern of such space frames supported by RC frames and establishing easy screening methods and design procedures, thereby preventing similar damages in the future.

7   Acknowledgements

This research was supported by MLIT and the Japan Building Disaster Prevention Association, Japan. The authors are also deeply thankful to Mr. Akira Jitsuishi of Taiyo Kogyo Co. Ltd., Dr. Yoshinao Konishi of Nippon Steel Engineering Co. Ltd., Dr. Tadashi Ishihara of Building Research Institute, Prof. Tetsuo Yamashita of Kogakuin University, and Prof. Satoshi Yamada of Tokyo Institute of Technology for providing building information and valuable advice.

References

Architecture Institute of Japan (2017) *AIJ Recommendation for Design of Latticed Shell Roof Structures* (in Japanese).
Architecture Institute of Japan (2018) *Report on 2016 Kumamoto Earthquake* (in Japanese).
Institute for Sophisticating Technique of Construction in Japan (2015) *Concepts for Seismic Diagnosis and Retrofit of Steel Roofs with RC Substructures*. Gihodo Shuppan (in Japanese).
Takeuchi, T. & Matsui, R. (2015) Cumulative deformation capacity of steel braces under various cyclic loading histories. *Journal of Structural Engineering*, ASCE, 141(7), 04014175-1–04014175-11.
Taniguchi, H., Kato, B., Nakamura, N., Saeki, T., Hirotani, T. & Aikawa, Y. (1994) Study on restoring force characteristics of X-shaped braced steel frames. *Journal of Structural Engineering*, AIJ, 37B, 303–316. (in Japanese).

Chapter 8

Nonlinear bifurcation behaviours of Incheon International Airport Terminal-2 by substitutional method of boundary rigidity

Seung-Deog Kim

I Introduction

Space frames with a single-layered grid have many more attractions than others because the system can be built economically with a dynamic shape. But sometimes that poses a little difficulty, especially in stability design of shell-like structures with a lot of calculations because of rigid joints which cannot be treated as hinges. Dome-shaped space frames have to be checked for snapping phenomena, such as snap-through and bifurcation in structural design. To check the overall buckling in design procedure, we need geometrical nonlinear analysis so that design load is not higher than the instability critical point for safety of the structure itself. If the structure consists of only rigid joints, the amount of calculation increases in geometrical progression.

In this study, we suggest a substitution method for simulation of bifurcation behaviours in the design procedure of Incheon International Airport Terminal-2, shown in Figure 8.1. The terminal has two parts of whole roof structure where the outside of the middle ring is double layered and the inside one is single layered. The nonlinear analysis of the whole roof results in too much iterative work. Therefore, we select only the inside part by introducing elastic springs at the boundary. The spring coefficients are obtained from a relation between the displacement of the inner ring at the original model and reaction of the ring as the boundary in the selected centre part. All geometrical nonlinear analyses are performed by NASS, which has been developed by the author as a multipurpose software for spatial structures and approved with many research achievements (Kim *et al.*, 1990, 1991, 1994, 1997, 2003; Kim et al., 2002; Kim and Kim, 2010).

2 Simplified test model

This section will show some numerical test results with simplified models to check whether the substitutional method to be suggested is reasonable or not, as shown in Figure 8.2. Figure 8.2a shows a simplified full model at 1/8 the scale of the original structure which has spans of 14.46 m in long direction and 8.35 m in short direction. Figure 8.2b is a core model to be chosen only inner ring part with fixed boundary which has spans of 7.07 m in length and 4.08 m in width. And Figure 8.2c is a spring model which has elastic springs in the x-, y-, and z-direction at all boundary nodes, respectively, and the boundary of the inner ring is substituted by elastic springs for movement of the boundary.

Figure 8.1 Incheon International Airport Terminal-2.

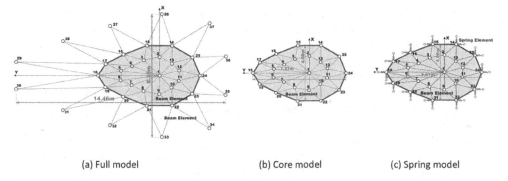

(a) Full model   (b) Core model   (c) Spring model

Figure 8.2 Simplified numerical test models.

Table 8.1 Comparison of displacements in full model

Node	Linear(cm)			Nonlinear(cm)			Error
	Mov-X	Mov-Y	Mov-Z	Mov-X	Mov-Y	Mov-Z	Mov-Z
1	0.000	−0.025	−0.661	0.000	−0.024	−0.699	5.80%
4	−0.124	0.000	−1.051	−0.126	0.002	−1.107	5.31%
5	−0.009	0.031	−1.225	−0.009	0.033	−1.287	5.07%
6	0.009	0.031	−1.225	0.009	0.033	−1.287	5.07%
7	0.124	0.000	−1.051	0.126	0.002	−1.107	5.31%
18	0.000	0.007	−1.156	0.000	0.007	−1.193	3.17%

To calculate the stiffness of elastic springs, we suggest the relation between the displacements of the inner ring in the full model and the reaction forces at the boundary of the core model using the equation $f = kd$, where f = reaction, d = displacement, and k = the spring coefficient. The main aim of this study is to reduce the amount of nonlinear calculation so that the boundary rigidity of springs is obtained by linear analysis.

First, the differences between linear and nonlinear analysis are compared in Table 8.1 and Table 8.2 for a full model and a core model, respectively. The full model showed around 5% and the core model around 3–4% difference.

Table 8.2 Comparison of displacements in core model

Node	Linear(cm)			Nonlinear(cm)			Error
	Mov-X	Mov-Y	Mov-Z	Mov-X	Mov-Y	Mov-Z	Mov-Z
I	0.000	0.061	−0.511	0.000	0.064	−0.532	4.29%
4	−0.051	0.042	−0.491	−0.052	0.044	−0.509	3.69%
5	−0.005	0.037	−0.453	−0.005	0.039	−0.470	3.55%

(a) Full model

(b) Core model

Figure 8.3 The first buckling modes.

Table 8.3 Ratio of design load to buckling load in full model (node I)

Imperfection	Buckling Load(A) (kgf)	Displacement (cm)	Design Load(B) (kgf)	Ratio (B/A)
–	2109.36	1.1	1383.235	65.58%
0.1%	2099.071	1.1	1383.235	65.90%
0.2%	2068.476	1.1	1383.235	66.87%

Table 8.4 Ratio of design load to buckling load in core model (node I)

Imperfection	Buckling Load(A) (kgf)	Displacement (cm)	Design Load(B) (kgf)	Ratio (B/A)
–	4885.734	2.15	1383.235	28.31%
0.1%	4931.109	2.2	1383.235	28.05%
0.2%	4873.571	2.2	1383.235	28.38%

For checking of instability phenomena, the buckling modes of each model are shown in Figure 8.3. These first buckling modes are used as initial imperfection for each model. The ratios of design load to buckling load are summarized for each model in Tables 8.3, 8.4, and 8.5, respectively. The full models show higher ratios around of 65–66% than the core models of around 28%. That is from the longer span of the full model, but with the same member size. If elastic springs are introduced at all boundary nodes in Table 8.5, the ratios become

Table 8.5 Ratio of design load to buckling load in spring model (node 1)

Revision	Semi Rigidity	Buckling Load(A) (kgf)	Displacement (cm)	Design Load(B) (kgf)	Ratio(B/A)
Nothing	100%	4409.817	5.6	1383.235	31.37%
1st	100%	4342.981	3	1383.235	31.85%
2nd	100%	4332.651	3.1	1383.235	31.93%
3rd	100%	4232.057	3.1	1383.235	32.68%

higher than the core model, from 28% to 31–32%. That is from the introduction of elastic springs, so that all boundaries become moveable a little more easily. In these results, we can see that there are no big differences even if we consider different initial imperfections, 0.1% and 0.2%, and different semi-rigidities of joints, 90% and 80%.

3   Incheon International Airport Terminal-2

The nonlinear instability analysis is applied to Incheon International Airport Terminal-2, which has spans of 141.7 m long and 80.1 m wide, and 60.76 m and 38.86 m of inner ring, respectively. The skeletons of the structure are shown in Figure 8.4. The outer part of the roof consists of a double-layered grid supported by some columns, and the inner part of a ring consists of a single layer. All member sizes are as follows;

Upper part: B-400 × 100 × 12
Lower part: H-390 × 300 × 10 × 16
Inner ring: H-300 × 300 × 10 × 15
Outer ring: H-200 × 200 × 8 × 12
Brace truss: H-200 × 100 × 5.5 × 8

The buckling modes of the roof are shown in Figure 8.5, which is used for initial imperfect shape in bifurcation analysis. Finite element modelling and layout of nodal numbers are

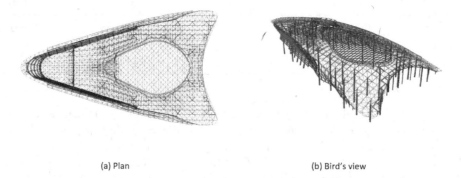

(a) Plan   (b) Bird's view

Figure 8.4 Skeletons of Incheon International Airport Terminal-2.

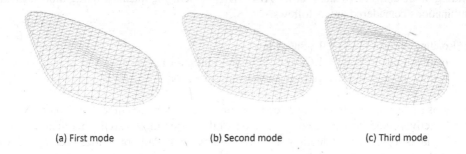

(a) First mode   (b) Second mode   (c) Third mode

Figure 8.5 Buckling modes of Incheon International Airport Terminal-2.

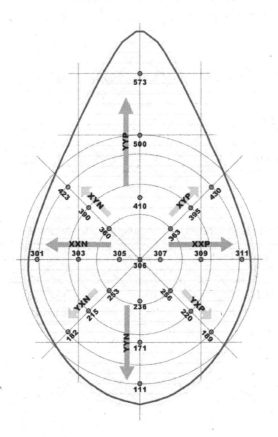

Figure 8.6 Modelling and layout of nodal numbers.

shown in Figure 8.6, and the grouping name such as XXP, XXN, YYP, and YYN are used in Figure 8.8.

To check nonlinear critical points, we consider some variables such as semi-rigidities: 100%, 90%, and 80%; initial imperfections: 0.0%, 0.1%, and 0.2%; buckling modes: 1st, 2nd, and 3rd; snow load distribution: uniform and eccentric. Therefore, the name of each model is "s00i0m0s0" which means semi-rigidity of inner joints, initial imperfection, buckling mode and snow load in order. Also, "SP" in the name means a spring model. Load combinations considered are as follows:

Loading mode 1: 1.2D + 1.6Su(uniform)
Loading mode 2: 1.2D + 1.6Sx(eccentric in x-direction)
Loading mode 3: 1.2D + 1.6Sy(eccentric in y-direction)

The ratios of design load to buckling load are summarized for each model in Tables 8.6 and 8.7, respectively, and we can see that the ratios of the spring model are higher than the core model. That is from the introduction of elastic springs, so that all boundaries become moveable a little easier. From the results, we can find that the ratios of design load to buckling load are less than 25% in all analytical models. Some selected figures of deformation shapes and nonlinear displacement curves are shown in Figures 8.7 and 8.8.

Table 8.6 Ratio of design load to buckling load in core models (node 306)

Model Name	Buckling Load (A) (kgf)	Displacement (cm)	Design Load (B) (kgf)	Ratio (B/A) (%)
IAsl0il m l su	17330.260	60	2492.748	14.38
IAsl0il m l sx	15436.730	57	1835.020	11.89

Table 8.7 Ratio of design load to buckling load in spring models (node 306)

Model Name	Buckling Load (A) (kgf)	Displacement (cm)	Design Load (B) (kgf)	Ratio (B/A) (%)
IAsl0il m l su_SP	10345.360	63	2492.748	24.10
IAsl0il m l sx_SP	10328.170	60	1835.020	17.77

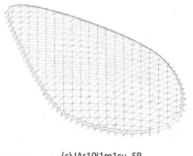
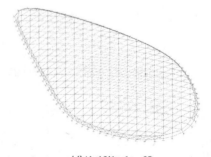

(c) IAs10i1m1su_SP   (d) IAs10i1m1sx_SP

Figure 8.7 Deformation shapes.

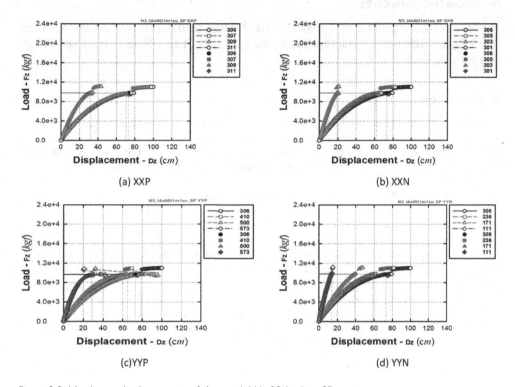

Figure 8.8 Nonlinear displacements of the model IAs08iImIsu_SP.

4   Conclusions

Space frames with single-layered grids have much more attractions than others because the system can be built with dynamic shape economically. But sometimes that has some difficulty, especially in stability design of shell-like structures with a lot of calculation works due to rigid joints. If the structure consists of only rigid joints, the amount of calculation amount increases in geometrical progression.

In this study, we have suggested a substitutional method for simulation of bifurcation behaviours in design procedure of Incheon International Airport Terminal-2, where the outer part of the roof consists of a double-layered grid supported by some columns, and the inner part of a ring consists of a single layer. As an analytical model, we selected only the inside part by introducing elastic springs at the boundary. The spring coefficients are obtained from a relation between displacement of the inner ring in the original full model and reaction of the ring as the boundary in the selected centre part.

We have considered some variables for analytical models, such as semi-rigidities (100%, 90%, and 80%), initial imperfections (0.0%, 0.1%, and 0.2%), buckling modes (1st, 2nd, and 3rd), and snow load distribution (uniform and eccentric).

Finally, we found that the ratios of design load to buckling load in the spring model are higher than the core model because all boundaries in the spring model became movable more easily. And the results showed that the buckling ratios are less than 25% in all analytical models.

Acknowledgements

This research was supported by a grant (19AUDP-B100343–05) from Architecture & Urban Development Research Program funded by the Ministry of Land, Infrastructure and Transport of the Korean government.

References

Kim, S.D. (2002) Nonlinear instability analysis of framed space structures with semi-rigid joint. *Journal of Architectural Institute of Korea*, 18(3), 52–62.

Kim, S.D. & Hangai, Y. (1991) Direct and indirect snappings of shallow E.P. shells under the up-and-down earthquake excitation. *Proceedings of International IASS Symposium 91*, Copenhagen, 3, 289–296.

Kim, S.D. & Kim, N.S. (2010) A study of the Snapping investigations of Seoul Southwest Baseball Dome. *Journal of the Korean Association for Spatial Structures*, 10(4), 133–140.

Kim, S.D., Tanami, T. & Hangai, Y. (1990) Direct and indirect snapping behaviors of shallow truss models. *Bulletin of ERS*, (23), 73–86.

Kim, S.D., Kang, M.M., Kwun, T.J. & Hangai, Y. (1994) Damping influence of simple shell-like shallow models to dynamic buckling. In: Papadrakakis, M. & Topping, B.H.V. (eds.) *Advances in Computational Mechanics*. Athens. pp. 99–105.

Kim, S.D., Kang, M.M., Kwun, T.J. & Hangai, Y. (1997) Dynamic instability of shell-like shallow trusses considering damping. *Computers & Structures*, 64(1–4, May), 481–489.

Kim, S.D., Kim, H.S. & Kang, M.M. (2003) A study of the nonlinear dynamic instability of hybrid cable dome structures. *Structural Engineering and Mechanics*, 15(6), 653–668.

Chapter 9

Roof covering effect on the structural behaviour of an annular crossed cable-truss structure

Suduo Xue, Renjie Liu, Xiongyan Li, and Marijke Mollaert

1 Introduction

The annular crossed cable-truss structure (ACCTS) has been proven to be a promising structure with the advantages of lightweight, feasible constructability, good structural stiffness, reasonable behaviour under asymmetric load cases, the possibility to be used in a large span, and satisfying ability to resist disproportionate collapse (Liu, 2014, 2017a, 2017b; Xue, 2017). However, the ACCTS alone cannot play a role as a complete roof of a building. A roof covering should be integrated with the ACCTS and work together to bear the loads. Due to the advantage of the low self-weight, the possibility to cover a large span, and various choices in shape, tensile membranes are widely used nowadays (Bridgens, 2004). Tensile membranes can act as a stand-alone structure and can also be integrated with other structures. However, the possibility of covering the ACCTS with a tensile membrane has not yet been verified. The roof covering effects on the static behaviour and on the ability to resisting disproportionate collapse should be discussed.

An integrated model which consists of a tensile membrane and the ACCTS is generated in this chapter to validate the possibility of covering the ACCTS with a tensile membrane. The size, materials, and cross-sections in the integrated model are consistent with the experimental setup which was built earlier. Form-finding of the integrated model is done to evaluate the ability to maintain the shape and keep reasonable pre-stress (Aboul-Nasr, 2015). The form-finding of the integrated model is done by using EasyForm based on the Force Density Methods (Schek, 1974). The stress in the tensile membrane for form-finding is discussed here.

Based on the integrated model, the roof covering effects on the static behaviour and the ability to resist disproportionate collapse are discussed. The load cases applied to the integrated model. The cable-rupture plans, which are the same as those in the experimental setup, were imposed on the integrated model. The displacements and tensions from the integrated model are compared with the experimental results to discuss the roof covering effects on the structural stiffness, the behaviour under asymmetric load cases, and the ability to resist disproportionate collapse.

A pure ACCTS with the same parameters as the experimental setup was also generated using EasyForm. The same load cases and cable-rupture plans were imposed on the pure ACCTS generated using EasyForm. Based on the pure ACCTS, one can correctly compare the structural behaviour of the structure with and without the tensile membrane. However, when the experimental setup was designed and built, the roof covering was not considered. The tensile membrane in the integrated model based on the experimental setup may be not perfect. It should be clarified that the purpose of designing the tensile membrane in this study is just to discuss the possibility of covering the ACCTS with a tensile membrane rather than to design the best tensile membrane for it, or even design an appropriate integrated system ACCTS with a membrane cover.

2   Roof covering effect on static behaviour

2.1   External load cases

The three load cases on the integrated model are illustrated in Figure 9.1. The same load cases are applied to the pure ACCTS using EasyForm in Figure 9.2. The concentrated loads were applied to all the bottom connections in the first load case. A half and a quarter of the bottom connections carrying the concentrated loads are used for the second load case and the third load case, respectively. The concentrated loads on the bottom connections were added by six load levels and reached 1.2 kN in the sixth load level. However, the concentrated loads on the bottom connections were 1.24 kN or 1.23 kN in Figure 9.1. The fact was that the values contain the self-weight of the structural elements which had been calculated automatically by the software.

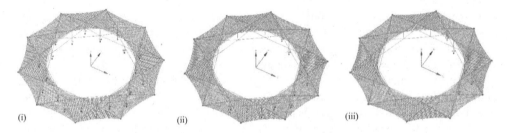

(i)   (ii)   (iii)

Figure 9.1 Load cases on integrated model: (i) the first load case, (ii) the second load case, and (iii) the third load case.

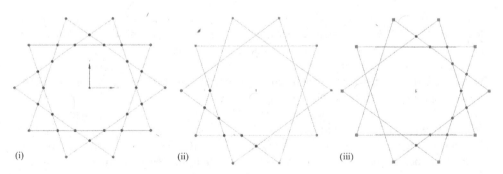

(i)   (ii)   (iii)

Figure 9.2 Load cases on pure ACCTS in EasyForm: (i) the first load case, (ii) the second load case, and (iii) the third load case.

2.2   Effect on structural stiffness

The roof covering effects on structural stiffness were discussed based on the displacements and tensions of the integrated model on the six load levels of the first load case. The load-displacement curves of different connections as well as the load-tension curves of different bottom cables were presented.

2.2.1   Discussion on displacements

In Figure 9.3, the size of the symbol on each connection represents the size of the displacement. The distribution of the displacements was uniform. The table on the left part of the figure shows that the maximum vertical displacement in the integrated model was −3 cm, which was 1.7% of the span. It indicates that the stiffness of the integrated model was still reasonable. However, the vertical displacements in the integrated model were slightly larger than the values of the corresponding connections in the experimental setup. However, the difference was so small that it could be neglected in view of the practical application.

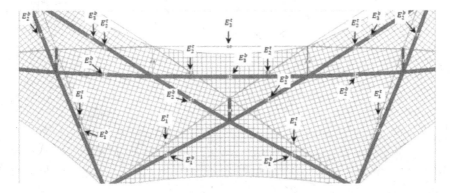

Figure 9.3 The forces of cables of the integrated model under symmetric load case (level 6) [kN].

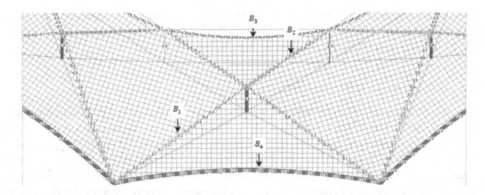

Figure 9.4 The tensions of boundary cables under symmetric load case (level 6) [kN].

Table 9.1 The tensions of the integrated model in Figure 9.3 and Figure 9.4 [kN]

Bottom cable	Tension	Top cable	Tension	Boundary cable	Tension
E_1^b	9.6	E_1^l	0.0	B_1	0.8
E_2^b	9.4	E_2^l	0.3	B_2	1.0
E_3^b	9.5	E_3^l	0.0	B_3	4.4
–	–	–	–	B_4	12.6

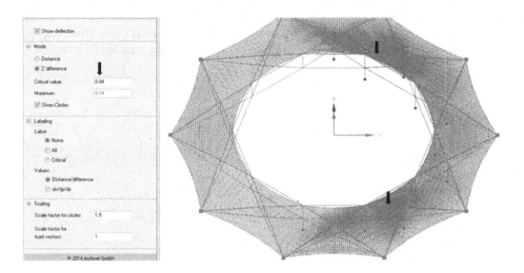

Figure 9.5 The distribution of vertical displacements under the second load case [m].

2.3   Effect on the behaviour under asymmetric load cases

The roof covering effects on the behaviour under asymmetric load cases are discussed based on the displacements and tensions of the integrated model, experimental setup, and the pure ACCTS in EasyForm under the second and third load cases.

2.3.1   Discussion on displacements

The vertical displacements in the integrated model under the second and third load cases are shown in Figure 9.5 and Figure 9.6, respectively. The maximum vertical displacement under both second and third load cases was 0.04 m, which was 2.3% of the span. The maximum vertical displacement was small and acceptable. In Figure 9.5 and Figure 9.6, the distribution of vertical displacements, which was about 0.04 m (the critical value emphasized in figures), was depicted. The unloaded zones of these load cases were not influenced by the loads. The vertical displacements of about 0.04 m were limited within the loaded zones.

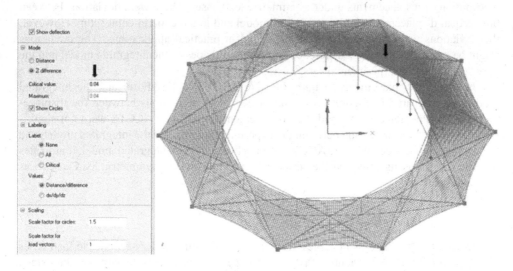

Figure 9.6 The distribution of vertical displacements under the third load case [m].

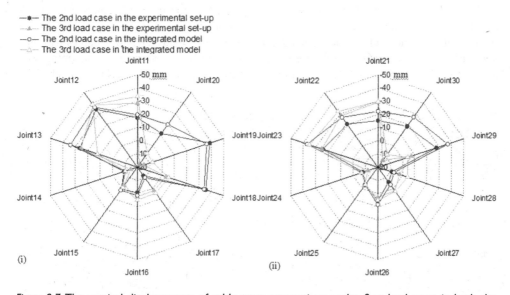

Figure 9.7 The vertical displacements of cable-strut connections under first load cases in both the experimental setup and the integrated model: (i) top connections of outer struts, (ii) top connections of inner struts.

The vertical displacements of the cable-strut connections under asymmetric load cases in both the experimental setup and the integrated model are depicted in Figure 9.7. The values of the integrated model and the pure ACCTS in EasyForm are compared. The numbers of joints in these figures are consistent with the number of testing arrangements. Most values from the integrated model agreed well with the experimental results. It indicates that the roof covering of the tensile membrane did not have an obvious effect on both values and the distribution of displacements under asymmetric load cases. There were deviations between the vertical displacements in the integrated model and in the experimental setup. However, the deviations were small and can be neglected in practical applications. The deviations might be caused by the slight difference between the shapes of the integrated model and the pure ACCTS in the experimental setup.

The vertical displacements in both the pure ACCTS in EasyForm and the integrated model are compared in Figure 9.8. It can be seen that derivations between the displacements in the integrated model and the displacements in the pure ACCTS were not significant. Based on the comparison between the experimental setup and the integrated model, and the comparison between the pure ACCTS in EasyForm and the integrated model, it indicates that the roof covering effect on the structural behaviour under asymmetric load cases was not significant.

2.3.2   Discussion on tensions

The distribution of tensions in cables in the integrated model under asymmetric load cases is shown in Figure 9.9. In Figure 9.9(i) and (iii), the tensions of bottom cables in the loaded zones were about 10 kN which was quite close to the values under the symmetric load case in Figure 9.2. The asymmetric cases did not cause a significant harmful effect on bottom cables in the loaded zones. Tensions of bottom cables in the unloaded zones were much smaller than those in the loaded zones. It indicates that the loads were mainly carried by the bottom cables in the loaded zones. The influence of external loads on bottom cables was limited within the loaded zones. In Figure 9.10, the tensions of bottom cables in the integrated model agreed well with the tensions in the experimental setup. Figure 9.11 shows that the

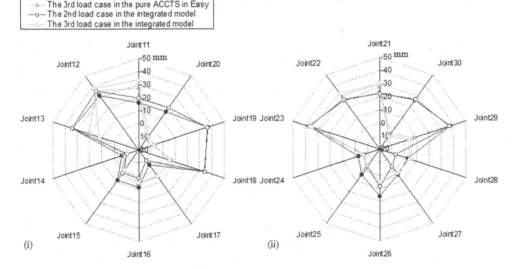

Figure 9.8 The vertical displacements of cable-strut connections under first load cases in both the pure ACCTS in EasyForm and the integrated model: (i) top connections of outer struts, (ii) top connections of inner struts.

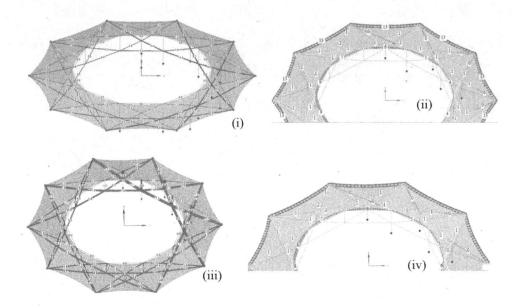

Figure 9.9 The distribution of tensions of cables in the integrated model under asymmetric load cases [kN]: (i) cables of the ACCTS under the second load case, (ii) boundary cables of the tensile membrane under the second load case, (iii) cables of the ACCTS under the third load case, and (iv) boundary cables of the tensile membrane under the third load case.

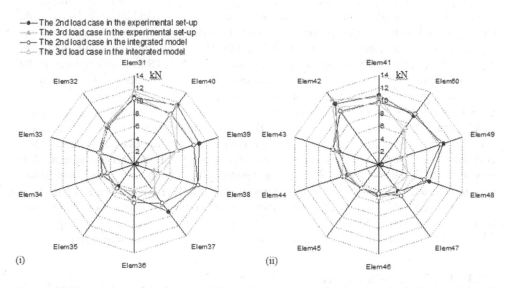

Figure 9.10 The tensions of the bottom cables under asymmetric load cases in both the experimental setup and the integrated model: (i) bottom-outermost cables and (ii) bottom-innermost cables.

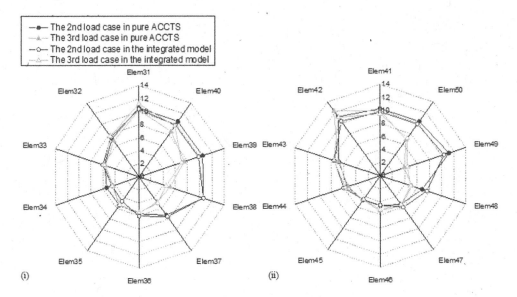

Figure 9.11 The tensions of the bottom cables under asymmetric load cases in both the pure ACCTS in EasyForm and the integrated model: (i) bottom-outermost cables and (ii) bottom-innermost cables.

tensions of bottom cables in the pure ACCTS in EasyForm have good agreement with the corresponding tensions in the integrated model.

In Figure 9.9(i) and (iii), top cables of the ACCTS in the unloaded zones were still in tension. Tensions of top cables in the loaded zones decreased a lot. Slacking was observed (the tension was 0 kN) in some top cables in the loaded zones under asymmetric load cases. The influence of external loads on top cables of the ACCTS was also limited within the loaded zones. However, the 'tensile action' of the top cables of the ACCTS had been replaced by the tensile membrane as well as the boundary cables of the tensile membrane. The slacking of several top cables of the ACCTS did not influence the behaviour under asymmetric load cases.

In Figure 9.9(ii) and (iv), a few boundary cables in the loaded zones between membrane panels became slack (the tension was 0 kN). But tensions of boundary cables in the unloaded zones between membrane panels did not change, compared to the values under symmetric load case. However, the boundary cables between membrane panels did not play an important role in tensioning the tensile membrane. The outermost and innermost boundary cables, which played the main role in tensioning the tensile membrane, were still well-tensioned. The tensions of outermost boundary cables did not change significantly, compared to the values under the symmetric load case. The tensions of the innermost boundary cables in the loaded zones ranged from 4 kN to 9 kN (from the centre of the loaded zone to the centre of the unloaded zone).

2.3.3   Discussion on the effect

Based on the vertical displacements and tensions under asymmetric load cases, the roof covering effect on the behaviour under asymmetric load cases can be summarized. The integrated model showed the same behaviour as the experimental setup and the pure ACCTS in

EasyForm to asymmetric load cases, the roof covering of a tensile membrane did not have an obvious effect on the behaviour under asymmetric load cases. The integrated model was sufficient to discuss the possibility of covering the ACCTS with a tensile membrane, and the roof covering effect. However, in the design of a practical application, the values of the pre-stress in the membrane and cables still have to be verified and increased in case wrinkling or ponding could occur.

3 Roof covering effect on the ability to resist disproportionate collapse

Three cable-rupture plans are imposed on the integrated model shown in Figure 9.12. The same cables are removed in the pure ACCTS in EasyForm. The cable-rupture plans are consistent with those performed on the experimental set-up. The displacements and tensions in the integrated model and in the pure ACCTS in EasyForm under the cable-rupture plans are obtained under the symmetric load case. The results are compared with the corresponding experimental data to discuss the roof covering effects on the ability to resist disproportionate collapse.

3.1 Results and discussion

3.1.1 Discussion on displacements

The distribution of the vertical displacements in the integrated model under cable-rupture plans is shown in Figure 9.12. The maximum vertical displacements under the P1, P2, and P3 were 0.07 m, 0.35 m, and 0.61 m, respectively. The maximum vertical displacements increased gradually with the number of the ruptured cables, rather than collapsing suddenly. Vertical displacements above the critical values (marked on the left part of the figure) were illustrated in the figure. The vertical displacements above the critical values were distributed within the local zones near the ruptured cable(s). The major part of the integrated model did not have significant displacements. It shows that the displacements in the integrated model under cable-rupture can be limited within the local zones, and the major part of the structure does not deform significantly.

Vertical displacements of cable-struts connections in both the experimental setup and the integrated model under three cable-rupture plans are displayed in Figure 9.13. Figure 9.14 shows that the vertical displacements of most connections in the integrated model agreed well with the experimental results and the results of the pure ACCTS in EasyForm. It indicates that the roof covering effect under cable rupture is not significant. However, obvious deviations existed in several singular joints, e.g. Joint 35 in Figure 9.14(iii), as well as Joint 35,

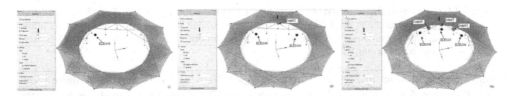

Figure 9.12 The distribution of the vertical displacements in the integrated model under cable-rupture plans [m]: (i) under P1, (ii) under P2, (iii) under P3.

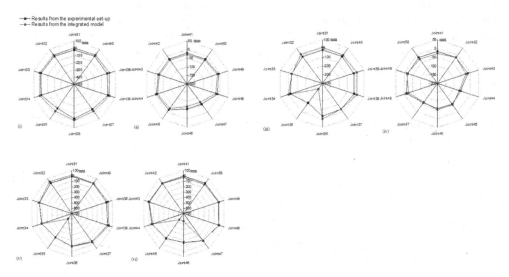

Figure 9.13 The vertical displacements of the integrated model and the experimental setup: (i) bottom outer struts under PI, (ii) bottom inner struts under PI, (iii) bottom outer struts under P2, (iv) bottom inner struts under P2, (v) bottom outer struts under P3, and (vi) bottom inner struts under P3.

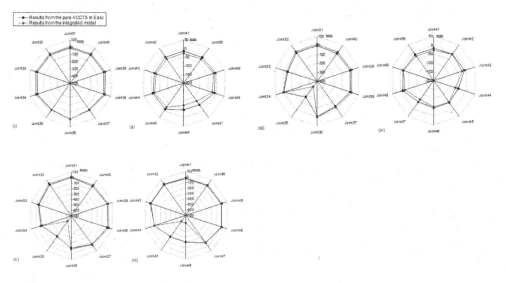

Figure 9.14 The vertical displacements of the integrated model and the pure ACCTS in Easy: (i) bottom outer struts under PI, (ii) bottom inner struts under PI, (iii) bottom outer struts under P2, (iv) bottom inner struts under P2, (v) bottom outer struts under P3, and (vi) bottom inner struts under P3.

Joint 45, and Joint 46 in Figures 9.14(v) and (vi). Vertical displacements of these joints in the integrated model were much larger than corresponding values in the experimental setup. The locations of these singular points can also be observed in Figure 9.12. It shows that these joints were located within the local zones near the rupture cables. The reason was the difference in the top layer between the integrated model and the experimental setup. The difference was that there were two cables between the connections to the integrated model.

One was the cable of the pure ACCTS and another one was the boundary cable in the tensile membrane.

3.1.2 Discussion on tensions

The distribution of the cables in the integrated model under cable-rupture plans is shown in Figure 9.15. The intact bottom cables which crossed with the ruptured cable(s) had significant larger tensions than the intact bottom cables which did not cross with the ruptured

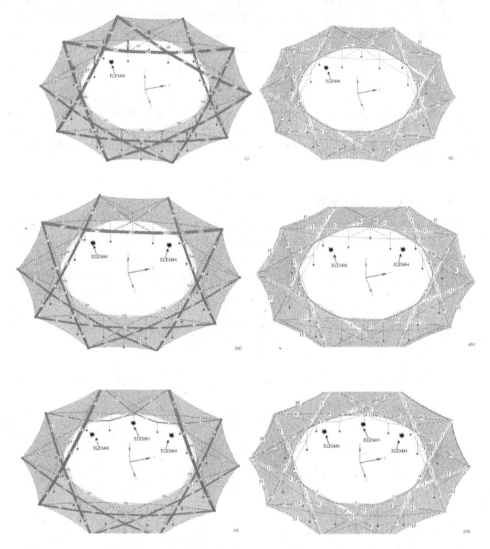

Figure 9.15 The distribution of tensions of cables in the integrated model under cable-rupture plans [kN]: (i) cables of the ACCTS under the P1, (ii) boundary cables of the tensile membrane under the P1, (iii) cables of the ACCTS under the P2, (iv) boundary cables of the tensile membrane under the P2, (v) cables of the ACCTS under the P3, and (vi) boundary cables of the tensile membrane under the P3.

cable(s). It indicates that the intact bottom cables which crossed with the ruptured cable(s) could act as alternative load paths, carry the loads instead of the ruptured cables, and limit the harmful effect within local zones. The maximum tension under the three cable-rupture plans was about 17 kN, which was still within the design axial force of the cable. No intact cable had the risk of breaking.

Figures 9.15(i), (iii), and (v) show that most of the top cables of the ACCTS in the integrated model were slack under all the three cable-rupture plans. However, the 'tensile action' by tensile membrane and the boundary cable of the tensile membrane had taken place. Especially, the innermost and the outermost boundary cables played the main role in tensioning the membrane. Figure 9.15 (ii), (iv), and (vi) show that tensions of the outermost boundary cables decreased slightly. Tensions of the innermost boundary cables in the local rupture zone decreased significantly. However, the innermost boundary cables out of the local rupture zone were still well-tensioned. It indicates that the major part of the integrated model was not influenced significantly.

Tensions of bottom cables in both the experimental set-up and the integrated model under cable-rupture plans are shown in Figure 9.16. Tensions of bottom cables from the integrated model agreed well with those from experimental set-up under all the three cable-rupture plans. It indicates that the roof covering did not influence the distribution as well as values of tensions of bottom cables.

3.2   Discussion on the effect

The roof covering effects on the ability to resist disproportionate collapse can be summarized as follows. The major part of the integrated model did not have significant displacements. It is found that the displacements in the integrated model under cable rupture can be limited within the local zones and the major part of the structure does not deform significantly.

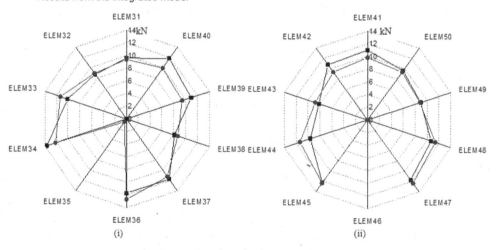

Figure 9.16 The tensions of bottom cables under cable-rupture plans: (i) outermost cables under the PI, (ii) innermost bottom cables under the PI.

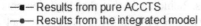

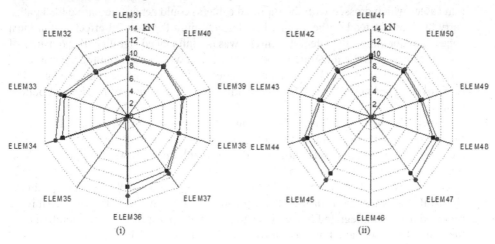

Figure 9.17 The tensions of bottom cables under cable-rupture plans: (i) outermost cables under the PI, (ii) innermost bottom cables under the PI.

Vertical displacements of most connections in the integrated model agreed well with the experimental results. It indicates that the displacements under cable rupture, with or without the roof covering, are not significant.

The intact bottom cables which crossed with the ruptured cable(s) could act as alternative load paths, carry the loads instead of the ruptured cables, and limit the harmful effect within local zones. A major part of the integrated model was not influenced significantly. Tensions of bottom cables from the integrated model agreed well with those from experimental setup under all three cable-rupture plans. The roof covering influenced neither the distribution nor the values of tensions of bottom cables.

4   Summary

An integrated model which consists of a tensile membrane and the ACCTS has been generated in this study to validate the possibility of covering the ACCTS with a tensile membrane.

- The roof covering effects on structural stiffness have been discussed based on displacements and tensions from the six load levels of the first load case. Firstly, the stiffness of the integrated model is found to be sufficient. Secondly, all the bottom cables in the integrated model are found to be as safe as those in the experimental setup. Thirdly, the top cables of the ACCTS can become slack under the sixth level of the symmetric load case.
- The roof covering effects on the ability to resist disproportionate collapse can be summarized as follows: The major part of the integrated model did not have significant displacements. It is found that the displacements in the integrated model under cable rupture can be limited within the local zones, and the major part of the structure does

not deform significantly. Vertical displacements of most connections in the integrated model agreed well with the experimental results. It indicates that the displacements under cable rupture with or without the roof covering are not significant. The intact bottom cables which crossed with the ruptured cable(s) could act as alternative load paths, carry the loads instead of the ruptured cables, and limit the harmful effect within local zones. A major part of the integrated model was not influenced significantly. Tensions of bottom cables from the integrated model agreed well with those from the experimental setup and the pure ACCTS in EasyForm under all the three cable-rupture plans. The roof covering did not influence the distribution as well as values of tensions of bottom cables.

• However, the tensile membrane designed in this study was not the best one. The curvature of the top layer was flat. In the design of a practical application, the curvatures of the top layer and bottom layer should be designed appropriately to take into consideration occurrence of wrinkling or ponding. There were redundant cables in the top layer since a top cable of the pure ACCTS and a boundary cable of the tensile membrane shared the same position. In fact, the top cables of the pure ACCTS can possibly be removed from the integrated model. How to design an appropriate tensile membrane for the integrated model and how to remove redundant cables will be the subject of further research.

References

Aboul-Nasr, G. & Mourad, S.A. (2015). An extended force density method for form finding of constrained cable nets. *Case Studies in Structural Engineering*, 3, 19–32.

Bridgens, B., Gosling, P. & Birchall, M.J.S. (2004) Tensile fabric structures: concepts, practice and developments. *The Structural Engineer*, 82(14), 21–27.

Liu, R., Xue, S., Sun, G. & Li, X. (2014) Formulas for the derivation of joint coordinates of annular crossed cable-truss structure in a pre-stressed state. *Journal of International Association for Shell and Spatial Structures*, 55(4), 223–228.

Liu, R., Li, X., Xue, S., Mollaert, M. & Ye, J. (2017a) Experimental and numerical research on Annular Crossed Cable-Truss Structure under cable rupture. *Earthquake Engineering and Engineering Vibration*, 16(3), 557–569.

Liu, R., Xue, S., Li, X., Mollaert, M. & Sun, G. (2017b) Preventing disproportionate displacements in an annular crossed cable-truss structure. *International Journal of Space Structures*, 32(1), 3–10.

Schek, H.J. (1974) The force density method for form finding and computation of general networks. *Computer Methods in Applied Mechanics and Engineering*, 3(1), 115–134.

Xue, S., Liu, R., Li, X. &Mollaert, M. (2017) Concept proposal and feasibility verification of the annular crossed cable-truss structure. *International Journal of Steel Structures*, 17(4), 1549–1560.

Chapter 10

Proposal of reusable kinetic structure – about the concept and structural characteristic for String Crescent Structure

Akira Tanaka and Masao Saitoh

1 Introduction

There have been many important themes of construction in the 21st century. In this chapter, we limit ourselves to the following:

- *Theme 1: Technology for measure resource depletion.* In recent years, the rate of material recycling has improved because of political and technological actions. Recycling is necessary for costly equipment used. The reuse rate for parts should be improved to avoid increasing costs.
- *Theme 2: Laboursaving.* In some countries like Japan, the shortage of skilled workers is a concern because of the ageing population. To prevent this, the use of construction technology for laboursaving is important.
- *Theme 3: Social effort for measuring climate change.* The increase in the number of disasters and disaster victims can be predicted through global warming. To prevent this, the development of temporary shelter space is important.

This chapter describes the proposal of the structure motivated by these three themes. Because it is used as a temporary space, this structure needs to have a balance between reuse performance and structural performance. First, a new concept of the reuse system is introduced.

1.1 Movable reuse system

Figure 10.1 provides an overview of the reuse system. The existing spatial structures, through the combination of "(A) Joint" and "(B) Construction," offer a diversity of structures.

To improve the reuse performance of the existing spatial structure, the emergence of "(C) Movable reuse system" is essential. "(C) Movable reuse system" is the reuse system whose structural geometry can be changed without replacement parts. The movable characteristics of the reuse system is affected by the choice of "(D) Movable element for structural geometry control" that was formed by the structural material, the part type, and the joint type. Briefly, the number of types of "(D) Movable element for structural geometry control" is three.

1.1.1 Change in joint position

Change in joint position refers to the shift elements for the joint point. This has been adopted as the pivot of String Scissor Structure (SSS) (Saitoh, 2007; Tanaka et al., 2007). By shifting the pivot position, the structural geometry changes to that of an arch such as flat arch,

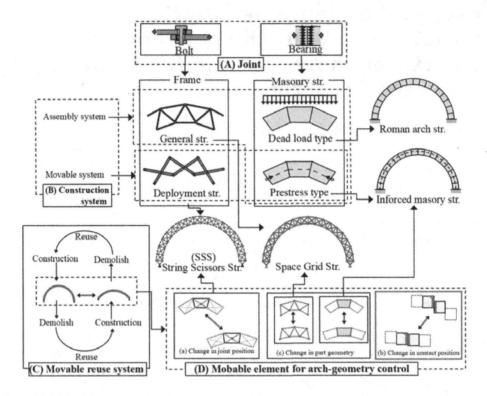

Figure 10.1 Movable reuse system from existing spatial structures.

semicircular arch, etc. (Figure 10.2). However, many pivots may lead to a deterioration of the mechanical performance of SSS.

1.1.2 Change in part geometry

Change in part geometry refers to two different elements. First, the element is the soft member such as a balloon; it is easy to force a deformation. However, this soft member has problems regarding improvement in rigidity. Another element is the replacement between members of different shapes. This has been adopted as the members of "sudare" and was designed by Toyo Ito. Various structural geometries consist of several members (Figure 10.3) (Ito et al., 2007; Ishida et al., 2006). However, the change in the shape of the structural surface creates the unnecessary members and the additional parts. This leads to an increase in costs of production and in storage space for unnecessary members. The larger this change area, the more the production cost increases.

1.1.3 Change in contact position

Change in contact position refers to the elements for sliding between contact surfaces such as the bone joint. This type leads to occurrence of the sliding deformation. The sliding deformation type is affected by the selection of the member shape (Figure 10.4). For example,

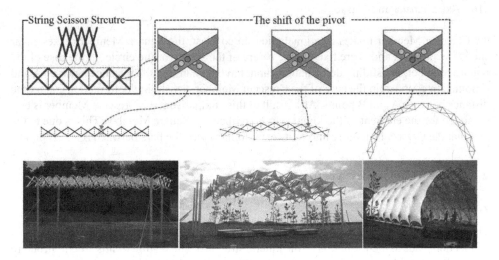

Figure 10.2 String scissor structure.

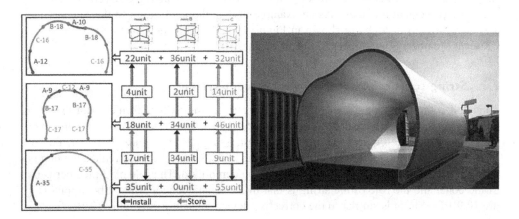

Figure 10.3 Sudare.

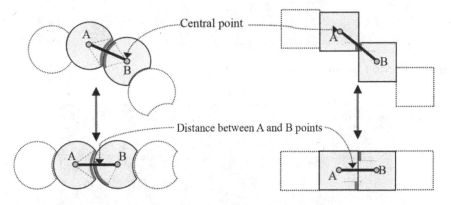

Figure 10.4 The sliding deformation.

the Crescent Member makes rotational slide; in contrast, the Square Member makes shear slide. The points A and B are fixed to the centre of the square or the circle. In the case of the Square Member, the sliding deformation cannot avoid changing the distance between A and B points. In contrast, in the case of the Crescent Member, the sliding deformation can fix the distance between A and B points. As a result of this comparison, the Crescent Member is better suited for the element of the kinetic structure than the Square Member. This is due to the fact that the Crescent Member makes it easier to introduce the bars between A and B points.

2   String Crescent Structure (SCS)

2.1   Crescent system

The bars between A and B points in Crescent Members are connected. The repetition of this connection leads to the formation of the kinetic structure. The feature of this kinetic structure is that the geometry changes like the snake (Figure 10.5). This kinetic structure is named the "Crescent system" (Tanaka, 2007). The Crescent system has the advantage of "Movable reuse system" by "Change in contact position," and it can change the structural geometry without replacement members. As an example of this, the application to the catenary arch and the semicircular arch are shown in Figure 10.6. This kinetic structure is formed with Crescent Members (C-member) and Arms (Figure 10.7).

2.2   String scissors structure from Crescent system

To stabilize the Crescent system, the strings are fixed and SCS is formed. But the fixing string is not adequate to ensure the stability of SCS, which therefore means that this problem-solving method requires the use of the prestressing string (Figure 10.8). The purpose of SCS is to serve as a temporary space (Theme 3). SCS is expected to reduce the number of used bolts, thus saving labour for construction/demolition. The role of C-member is that of the bone, and the role of the string is that of the ligament supporting a bone joint (Figure 10.9). The string is similar to the corrective device for scoliosis (Figure 10.10)

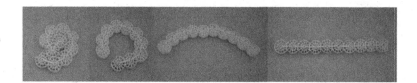

Figure 10.5 Crescent system.

Figure 10.6 Catenary arch and semicircular arch.

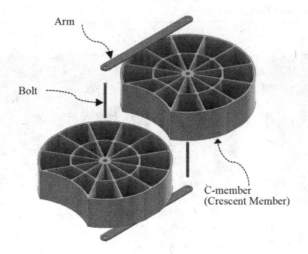

Figure 10.7 The member of construction.

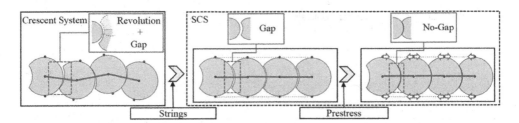

Figure 10.8 SCS from Crescent system.

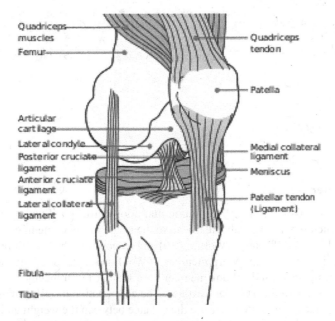

Figure 10.9 Ligament (Anterior Cruciate Ligament, 2019).

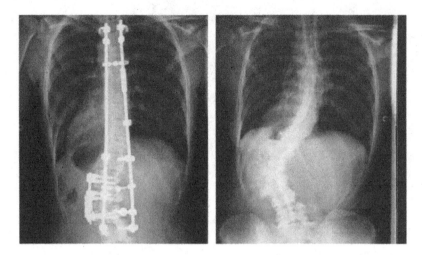

Figure 10.10 Corrective device for scoliosis.

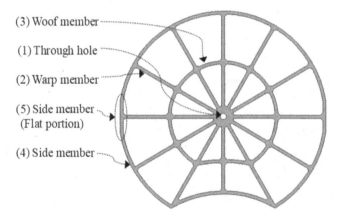

Figure 10.11 Detail of C-member.

2.3 C-member

The material of the C-member will be aluminium alloy extrusion (Figure 10.11). The hint for the bone structure with the trabecular lead to the shape of the C-member. The trabecular pattern is based on "Wolff's law" (Wolff, 2010) and responds to the flow of principal stress. The C-member is formed with "Warp member," "Woof member," and "Side member." The roles of "Warp member" and "Woof member" are that of the trabecular, and the role of "Side member" is that of the articular cartilage. The combination of "Warp member," "Woof member," and "Side member" will ensure the balance between the weight and the structural performance.

"(1) Through hole" is used for the bolt joint of the C-member and Arm. "(2) Warp member" refers to a reinforcing member of the side members. "(3) Woof member" is used for the buckling failure of Warp members. "(4) Side member" is for stress transfer mechanism between C-members, which is used for the joint between the string and the finishing material. The shape of the side member is that of a curved surface. As a result, a nonstandard bracket is necessary, albeit at a higher cost. To prevent this, "(5) Flat portion" is designed (Figure 10.12).

The C-member appears similar to the Swedish Cope of the log house, which differs from the fixing method as shown in Figure 10.13. Swedish Cope fixing elements are the joining joggle and the anchor bolt, which are hidden from the outside. The C-member fixing element is the string, which is fixed to the outside. For this reason, this part must be concerned about the appearances.

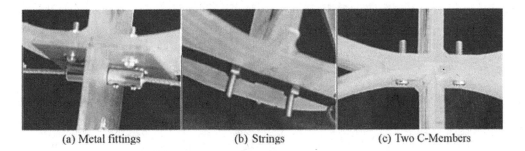

| (a) Metal fittings | (b) Strings | (c) Two C-Members |

Figure 10.12 The joining method of C-members.

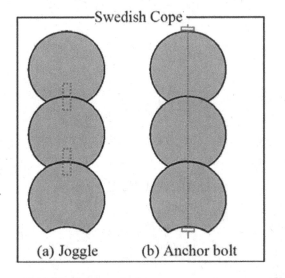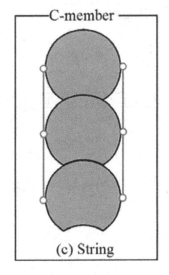

Swedish Cope   C-member

| (a) Joggle | (b) Anchor bolt | (c) String |

Figure 10.13 The fixing methods.

2.4   Arm

For manufacturing of the Crescent system, C-members are connected by Arms and many bolts at a factory. As a result, the number of processes of bolt joining in the outdoor construction site is reduced, and the construction period is shortened. In order to be carried from the factory to the outdoor construction site, the Crescent system should be folded compactly. To ensure this mobility, a loose hole between the bolt and bolt hole of the Arm should be designed (Figure 10.14). If the loose hole of the Arm is large, the derailment occurs between C-members, and the Crescent system becomes unstable. In contrast, if the Arm does not have the loose hole, the Crescent system does not become the kinetic structure, because there is a strong contact between C-members. Therefore, to make the stable kinetic structure, the proper size of the loose hole has to be designed (Figure 10.15).

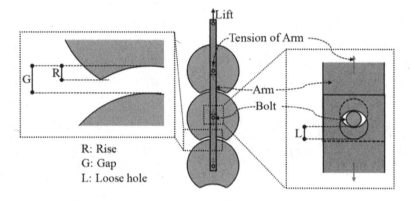

Figure 10.14 The size of gap and loose hole.

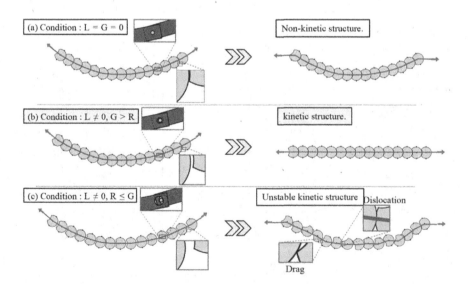

Figure 10.15 The role of the Arm in the Crescent system.

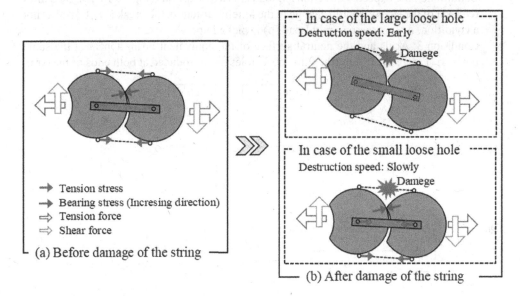

In case of the large loose hole
Destruction speed: Early
Damege

In case of the small loose hole
Destruction speed: Slowly
Damege

→ Tension stress
→ Bearing stress (Incresing direction)
⇒ Tension force
⇒ Shear force

(a) Before damage of the string

(b) After damage of the string

Figure 10.16 The role of the Arm in SCS.

In the case of SCS under the tension and the shear, when the string does not become damaged, the mechanical role of the Arm is negligible (Figure 10.16a). When the string is damaged, it has the risk of SCS becoming the unstable structure (Figure 10.16b). If the loose hole is large, C-members cannot contact each other. This phenomenon leads to sudden destruction of SCS and does not give evacuation time. To avoid this problem, the loose hole size should be small. This small hole enables the Arm to control the separation between C-members. This role of the Arm is the safety device of SCS.

3   Mechanical characteristics of SCS

3.1   The stress transfer mechanism between C-members

The resistance of SCS to wind must be considered, and the basic structure characteristic of SCS must be understood. To study this, a column geometry model is designed, and the stress transfer mechanism between C-members should be estimated.

3.1.1   The conditions for the estimation of the stress transfer mechanism

The three conditions for this estimation are described as follows:

- Condition 1: two different contact stresses occur between C-members (Figure 10.17). However, the involved stress is only the bearing stress. The reason for this will be described in section 4.1, "The material of experimental C-member," and the analysis of friction is difficult because the conditions change the gap size between C-members.

- Condition 2: the expected directions of load are two directions (Figure 10.18). SCS after removal of the strings should change to the kinetic structure, the breakage of Arms is not a condition. At this reason, only Load (b) is picked up.
- Condition 3: Arms and the neutral surface of the equivalent beam appear at the same position (Figure 10.19). Further, the loose holes are introduced at both ends of the Arm.

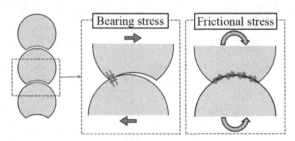

Figure 10.17 Two stresses between C-members.

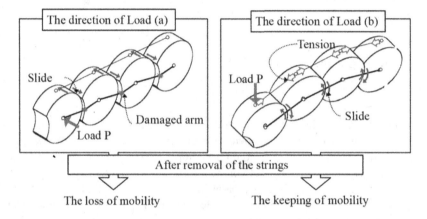

Figure 10.18 Deformation of SCS.

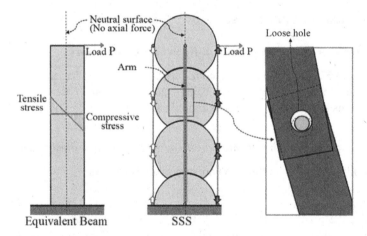

Figure 10.19 Arms and the neutral surface.

As this result, it can be considered that the axial force does not occur at the Arm under Load (b) that is shown in Figure 10.18, and the Arm's structural role can be ignored.

3.1.2 The estimation of the stress transfer mechanism

Based on the three conditions described in section 3.1.1, the estimation of the stress transfer mechanism between C-members is attempted (Figure 10.20). To secure the stability of SCS, the stress transfer mechanism changes because of the effects of the separation between C-members and the slackness of the strings.

3.1.2.1 THE STRESS TRANSFER MECHANISM UNDER SHEAR FORCE

The shear force affects the shear sliding and the separation between C-members, but it does not affect the axial force of the strings. Before the separation, the bearing stress decreases on the separation side and increases on the contact side. At the same time, the tension in the string is not generated. After the separation, the bearing stress increases on the contact side. At the same time, to prevent further separation, the low-tension stress is generated in the string.

3.1.2.2 THE STRESS TRANSFER MECHANISM UNDER BENDING MOMENT

The bending moment affects the rotational sliding between C-members and the axial force of the strings but does not affect the separation. Before the slackness in the string (the compression side), the tension in the string increases on the tension side but decreases on the compression side. After the slackness in the string (the compression side), the tension in the string (the tension side) and the bearing stress increase.

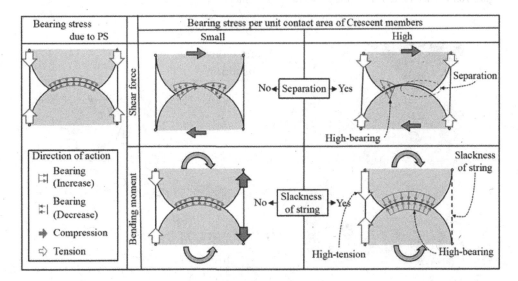

Figure 10.20 Stress transfer mechanism in Joint between C-members.

4   Load experiment

4.1   The material of experimental C-member

To reduce experimental costs, the epoxy resin was used in place of aluminium alloy. The fabrication method for epoxy C-member is described here and illustrated in Figure 10.21. (1) The molding box was shaped using a 3D printer. (2) The epoxy resin was poured into the molding box. (3) After vitrification, the molding box was removed. As a result, the unevenness such as a striped pattern is formed on the surface of C-member and makes a small gap at the contact portion between C-members (Figure 10.22). For this reason, the slide in the curved direction easily occurs. In contrast, the slide in the plane direction does not occur. These slides contribute greatly to the occurrence of sliding deformation under "Load (b)" (Figure 10.18).

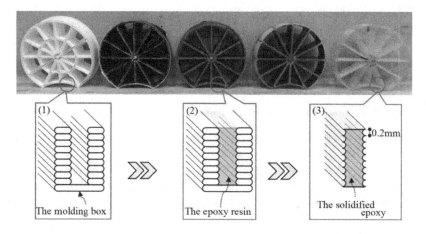

Figure 10.21 The fabrication method for epoxy C-member.

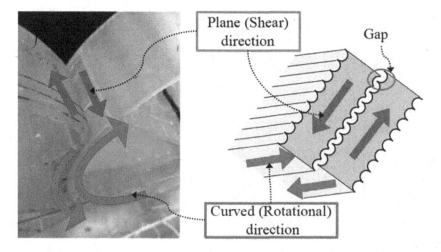

Figure 10.22 Two slide directions.

4.2   The purpose of load experiment

The number of C-members was three. To set the column shape of SCS, they were piled up on the semi-circular member (Figure 10.23a). The purpose of the load experiment is as follows:

- Load Experiment A: the mechanical comparison between different hollow shapes of C-members.
- Load Experiment B: the effect of prestress on the mechanical performance of SCS.

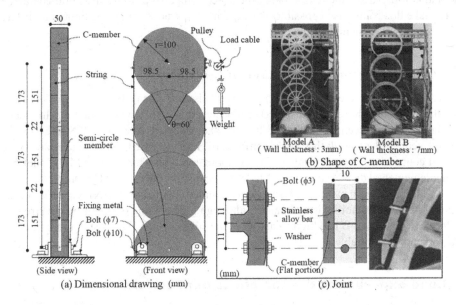

Figure 10.23   Load Experiment A.

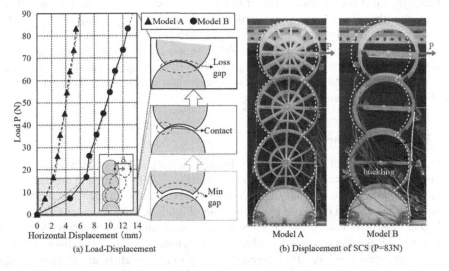

Figure 10.24   The result of Load Experiment A.

5   Load experiment A – comparison between different hollow shapes of C-members

5.1   The overview of load experimental A

Two types of C-members that are different hollow shape were used for Load Experiment A. The weights of both Models A and B were 360 g (Figure 10.23b). The material for the string was stainless alloy (Figure 10.23c). The condition of design for the string was non-loose hole and non-prestress, the reason being the sheer comparison of the structural characteristic. To make this comparison possible, the experiment parameter must be limited to different hollow shapes of C-members.

5.2   The effect of the hollow shape of C-members

The results of Load Experiment A are as follows (Figure 10.24a): Before the load value was more than 20N, high horizontal displacement occurred in both Models A and B. This cause is the minimal gap between C-members. After the load value was more than 20N, the incremental displacement decreased; that is, the stiffness of SCS increased. These causes are the loss gap and contact between C-members. When compared with Models A and B, Model A has smaller incremental displacement than Model B because the stiffness of C-members improved by the addition of warp members. Displacement under the load (P = 83N) is shown in Figure 10.24b. The overall stability is maintained, although the buckling of the string occurred.

Both Models A and B had high variation in the stress values. As a result, these numerical data were not quantifiable.

6   Load experiment B – the effect of prestress

6.1   The overview of load experimental B

Load Experiment B is performed in order to study the effect of prestress. The parts that are different from the case of Load Experiment A are the materials of string and the joint. The joint with the strand cables and the prestressing devices are shown in Figure 10.25. The trays are hung from the bottom edge of the left and right strand cables. As the prestressing method, the weights are put on the tray. The value of this weight is equivalent to the prestressing value. After achieving the target prestressing value – that is 10N, 35N, 70N, or 100N – the strand cable was fixed using a circular sleeve.

6.2   The effect of the prestress

Load Experiment B was able to demonstrate the effect of the prestress on SCS (Figure 10.26a). Prestress leads to an increase in the compression stress of C-member, and the difference in the role of compression stress under a low load (P < 40N). But the difference in this compression stress is lost under a high load exceeding 40N (Figure 10.26b). In case of the strings on the tension side, there is a difference in the magnitude of tension under a low load. But the difference in this tension is lost under a high load (Figures 10.26c, e). The distance between measurement points of Figure 10.26b and the string to above left is small. As a result, the loads when the effect of prestress is lost are the same (Figures 10.26b, c). In case of the string to above right, tensile stress of the string decreases with increasing load (Figure 10.26d).

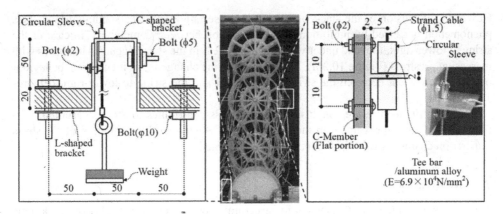

Figure 10.25 Load Experiment B.

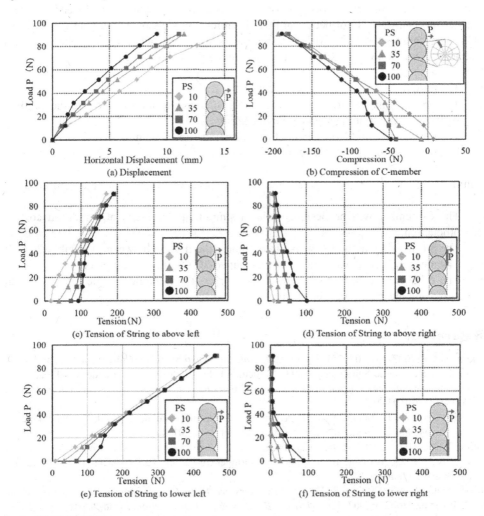

Figure 10.26 PS-effect.

However, the tension of the slackened string increases slightly under maximum load. This phenomenon is caused by inhibition of the separation between C-members. In case of the string to lower right, the load under which the slackness in the string occurred increases with increasing prestress (Figure 10.26f). The load where the change in the mechanical behaviour occurs is consistent between the string to lower left and the string to lower right (Figures 10.26e, f).

These results lead to the conclusion that prestress contributes to the displacement suppression and the improvement in the stability of SCS and can the estimation of the stress transfer mechanism shown in Figure 10.20 is confirmed.

7   Summary

The concept of "(D) Movable element for structural geometry control" leads to a new kinetic structure named the Crescent system. To ensure the stability of the Crescent system, SCS with the prestressing strings was proposed. SCS is a novel structural type where the stress transfer mechanism has not been clarified. Therefore, models of the column geometry of SCS were constructed and two load experiments were conducted to study the structural characteristics. These load experiments provided the following insights:

- The design of C-member had an effect on the stiffness of SCS.
- The prestress in the string enabled a stable SCS and an improvement to the bending stiffness and restraining slackness in the string.
- The prestress did not have an effect on stress under maximum load.
- Estimation of the stress transfer mechanism was supported.

Future themes that point to the practical use of SCS as a temporary space are as follows:

- The C-member will be designed with a simplified shape and at a reduced cost of production.
- The prestressing method will take into account the friction between C-members.
- The mechanical characteristic of an arc-shaped SCS will be investigated.
- The production cost performance of C-member; for example, simple cross-sectional shape and the use of inexpensive material.

References

Akira Tanaka (2007) No. 6078251. *The Building Unit, the Structure and the Construction*. Japan-Patent.
Akira Tanaka & Masao Saitoh, et al. (2007) Structural behaviour of vault roof under wind loads considering construction. *Journal of Structural and Construction Engineering (Transactions of AIJ)*, (611, January), 95–102.
Anterior Cruciate Ligament (2019) *Wikipedia-English*. [Online] Available from: https://en.wikipedia.org/wiki/Medial_collateral_ligament [Accessed 6 February 2019].
Julius Wolff, MD. (2010) *On the Inner Architecture of Bones and Its Importance for Bone Growth 468, 1056–1065*. Available from: https://link.springer.com/article/10.1007%2Fs11999-010-1239-2 [Accessed 10 January 2019].
Masao Saitoh. (2007) P2002–309674A. *String Scissors Structure & Constructional Element*. Japan-Patent.
Toyo Ito, et al. (2007) P2007–120207A. *The Building & the Construction*. Japan-Patent.
Yasuo Ishida, et al. (2006) *The Space of Aluminum*. pp. 126–127. ISBN-13:978-4786901898 Shinken-chiku-sha Co., Ltd, Tokyo, Japan.

Chapter 11

Minimal surface search by the tunnel method on the Plateau problem[1]

Toku Nishimura

1 Introduction

Membrane structures, constructed by lightweight and processed woven fabric material, are one example of tensile structures. Creep and stress relaxation occur easily in membrane materials. If the distribution of tensile stress on the membrane surface is unbalanced during the introduction of initial tensile stress, the stress distribution tends to transit to the lowest energy level and uniform. As a result, ponding and tear of membranes occur due to wrinkling and sagging. Above the mentioned, initial surface, everywhere uniformly stressed in membrane surface, is desirable to avoid winkling (Otto, 1969).

It is known that the uniform stress surface implies the least strain energy surface and corresponds to minimal area surface. The problem, which studies minimal surfaces spanned in boundaries, is called the *Plateau problem* (Hildebrant and Tromba, 1985). The problem is generally a multimodal nonlinear problem with some extremums. The sequential computation to minimize the surface area yields the coordinates of surfaces to the prescribed reducing area. This process leads to an optimization to minimize the surface area. Searching at any initial surfaces, the obtained minimal surface may be a local optimal solution and not always a global optimal solution. A numerical strategy is necessary to avoid capturing local minimal solutions and to obtain the least area surface.

Hinata *et al.* (1974) proposed a method to solve the Plateau problem by the finite element method. The following surface area functional J (Courant, 1950) by a single-valued function $f(x,y)$ is used:

$$J(x, y) = \iint \sqrt{1 + \left(\frac{\partial f}{\partial x}\right)^2 + \left(\frac{\partial f}{\partial y}\right)^2} \, dxdy \tag{1}$$

On the stationary condition $\delta J = 0$, the solution adopted the generalized Newton method. The numerical results are presented for Catenoid, the Courant problem, and so on. Especially in the relations between parameters (height and the coordinate of a saddle point) of the Courant problem, the existence of hysteresis has been pointed out. Ishihara and Ohmori (1993) defined a functional linearly combined with some other functional. To control the parameter consists of the function, the method, which are able to avoid singular points during the solution, was presented. Using the functional with constrained conditions concerning the volume and the area of surfaces, two minimal surfaces and one maximum surface were obtained in

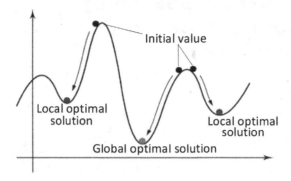

Figure 11.1 Initial dependence of nonlinear problem with multimodality.

the Courant problem. Tsutiya (1986) presented two methods where an area functional A and an energy functional E are stationary. These functionals are as follows:

$$A(f) = \iint \sqrt{\left|\frac{\partial f}{\partial x}\right|^2 \left|\frac{\partial f}{\partial y}\right|^2 + \left(\frac{\partial f}{\partial x} \cdot \frac{\partial f}{\partial y}\right)} \, dxdy \tag{2}$$

$$E(f) = \frac{1}{2} \iint \sqrt{\left|\frac{\partial f}{\partial x}\right|^2 + \left|\frac{\partial f}{\partial y}\right|^2} \, dxdy \tag{3}$$

The solution used the usual relaxation method. The numerical results are demonstrated on Catenoid, Courant problem, various knots, and conformal maps. The previous studies, however, do not mention how to find not only the local solution but the global optimal solution.

Solving the problems with some minimal surfaces by linearized incremental analysis, depending on the initial surface, the obtained minimal surface is not necessarily the global optimal solution, which is called least area surface. It is caused by initial independence on the nonlinear problem with multimodality, as shown in Figure 11.1. As a solution to the high initial independent optimal problem, recently heuristic methods such as generic algorithm, cellular automaton, simulated annealing, etc., have been applied in engineering fields. Nishimura and Yamanaka (2014) proposed a membrane shape analysis applying simulated annealing. The numerical results show that the global optimal solution can be found on both Catenoid and the Courant problem.

The tunnel method has been utilized for global optimization in multimodal problems. In this chapter, the algorithm using the tunnel method is presented to determine minimal area surfaces.

2   The tunnel method

The tunnel method is used to minimize the functions expressed in continuous variables. This method was put forward by Levy *et al.* (1982) and efficiently searches global minimum or maximum points of continuous functions with some extreme points.

In the tunnel method, sequential minimizing step and tunnelling step result in a global optimal solution.

2.1   Minimizing step

Setting an arbitrary initial value $x_1{}^0$ of function $f(x)$, we find the minimal solution $x_1{}^*$ by local search. In the local search, as the variation Δf of the function coincides with required value, the variable x is calculated by linearized incremental analysis. The basic equation is expressed as

$$f(x) = f(x_1{}^0) + \Delta f \approx f(x_1{}^0) + \left.\frac{\partial f}{\partial x}\right|_{x=x_1{}^0} \Delta x \tag{4}$$

The solution of Eq. (4) by the steepest descent method can be written as

$$\Delta x = \left.\frac{\partial f}{\partial x}^{-}\right|_{x=x_1{}^0} \Delta f \tag{5}$$

Here, $\left.\dfrac{\partial f}{\partial x}^{-}\right|_{x=x_1{}^0}$ is the generalized inverse of the gradient for the function. When the

gradient $\left.\dfrac{\partial f}{\partial x}^{-}\right|_{x=x_1{}^0}$ does not exist, x attains to a minimal solution.

2.2   Tunnelling step

We define the following tunnelling function:

$$T(x) = \frac{f(x) - f(x^*)}{\left\{(x-x^*)^T(x-x^*)\right\}^{\lambda}} \tag{6}$$

where x^* is a minimal solution in the former minimizing step. The power number λ of denominator means the intensity of the pole. Let us find the initial value $x_{k+1}{}^0$ for the next minimizing step, which satisfies Eq. (7).

$$f(x_{k+1}{}^0) \leq f(x_k{}^*), \quad x_{k+1}{}^0 \neq x_k{}^* \tag{7}$$

In the tunnelling step, when the next initial value $x_{k+1}{}^0$, which satisfies the following relation

$$f(x_{k+1}{}^0) \leq f(x_k{}^*) \tag{8}$$

cannot be found, we stop the search and consider the minimal solution $x_k{}^*$ as a global minimal solution. The schema of the tunnel method is shown in Figure 11.2.

Levy et al. (1982) introduced the idea of a pole where the tunnel function $T \neq 0$ at a minimal point $x = x_k{}^*$ and to get descent characteristics during the tunnelling step. Searching x with $T \leq 0$, the simultaneous nonlinear equations with n variables, which consists of $n-1$ artificial constraint conditions, is solved by the Newton method.

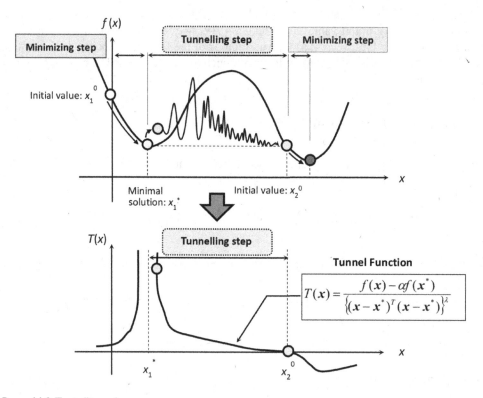

Figure 11.2 Tunnelling scheme.

In this membrane shape analysis, let x and the function $f(x)$ be the free nodal coordinates on the surface and the boundaries and surface area function $S(x)$, respectively. Eq. (5) is rewritten as:

$$\Delta x = \left.\frac{\partial S}{\partial x}^{-}\right|_{x=x_0} \left(S(x) - S(x_1^0)\right) = \left. a^{-}\right|_{x=x_0} \Delta S \tag{9}$$

where $a = \left.\dfrac{\partial S}{\partial x}\right|_{x=x_0}$ is the gradient vector of the surface area function $S(x)$, ΔS is the prescribed increment of the surface area, and a is the n row vector. The generalized inverse matrix $a^{-} = \left.\dfrac{\partial S}{\partial x}^{-}\right|_{x=x_0}$ is expressed as follows:

$$a^{-} = \frac{a^T}{aa^T}(a \neq 0), \ \ a^{-} = 0(a = 0) \tag{10}$$

Changing from minimizing step to tunnelling step, the coordinates x are shifted to $x^* + dx$ in the vicinity of a minimal point. Let dx be

$$dx = r \times rand, \quad -1.0 \leq rand \leq 1.0 \tag{11}$$

where r is a numerical parameter to shift the coordinates, and $rand$ is the uniform random number.

In the tunnelling step, we use the altered tunnelling function $T(x)$ as follows:

$$T(x) = \frac{S(x) - \alpha S(x^*)}{\left\{ (x - x^*)^T (x - x^*) \right\}^{\lambda}}, \quad 0 \leq \alpha \leq 1 \tag{12}$$

The numerator α is the control parameter. Linearizing Eq. (12), the following equation is iteratively solved until $T(x) \leq 0$.

$$T(x) = T(x_0) + \Delta T \approx T(x_0) + \left. \frac{\partial T}{\partial x} \right|_{x=x_0} \Delta x = 0 \tag{13}$$

where

$$\left. \frac{\partial T}{\partial x} \right|_{x=x_0} = b_1 + b_2 \tag{14}$$

$$b_1 = \frac{1}{\left\{ (x_0 - x^*)^T (x_0 - x^*) \right\}^{\lambda}} \left. \frac{\partial \left(S(x) - \alpha S(x^*) \right)}{\partial x} \right|_{x=x_0} = \frac{a|_{x=x_0}}{\left\{ (x_0 - x^*)^T (x_0 - x^*) \right\}^{\lambda}} \tag{15}$$

$$b_2 = \left(S(x_0) - \alpha S(x^*) \right) \cdot \left. \frac{\partial}{\partial x} \frac{1}{\left\{ (x - x^*)^T (x - x^*) \right\}^{\lambda+1}} \right|_{x=x_0}$$

$$= -2\lambda \frac{S(x_0) - \alpha S(x^*)}{\left\{ (x_0 - x^*)^T (x_0 - x^*) \right\}^{\lambda+1}} (x_0 - x^*) \tag{16}$$

and b_1 and b_2 are the n row vectors.
Using the steep descent method, the solution of Eq. (13) can be written as

$$\Delta x = -(b_1 + b_2)^- T(x_0) \tag{17}$$

The aforementioned computational flow is shown in Figure 11.3.

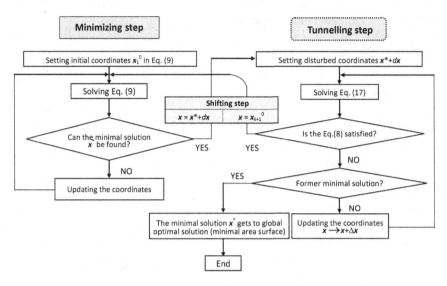

Figure 11.3 Tunnelling algorithm.

3   Implementation and numerical results

3.1   Courant problem

This problem is examined on spanned minimal area surfaces in a fixed boundary, which is a closed curve connecting two parallel arcs with perpendicular lines. The feature of this problem has been known as a multimodal problem which underlies a maximum and two minimal solutions.

The analysis model is symmetrical with the x-axis as shown in a model in Figure 11.4. The initial shape is cylindrical and divided by 20 circumferentially and by 10 vertically using triangular mesh. The radius R = 1.0, the height L = 1.5, and the angle of arc $\phi = 5\pi/6$. Numerical parameters are given in Table 11.1. The history of membrane surface area is shown in Figure 11.5. We employ the parameters as $\alpha = 0.97$, $\lambda = 1.0$ and $r = 0.01$ as shown by the shaded cell in Table 11.1. In Eq. (9), during the minimizing step, the increment of surface area ΔS is set as $- S/2000$ where S is the current surface area. Solving Eq. (9) from initial shape at A, the inner product $a^T a$ decreases less than 1.0×10^{-5} at B (437 steps). Since a^- vanishes, we get a minimal surface, whose area is 3.52622. Subsequently, the surface is varied randomly in the neighbourhood point C (438 steps). Afterward, the surface area is fluctuating and increasing until it is near 500 steps. Figure 11.6 shows the tunnelling function in Eq. (12) during the tunnelling step (438–763 steps). The initial value is 17.8774 at C. In subsequent steps, it confirms that the function repeats a slight fluctuation; however, the function shows an inclination to decrease asymptotically. At point D (763 steps), the surface area $S = 3.51674$, and it is less than the first minimal surface area. Accordingly, the numerical step is switched from tunnelling to minimizing. The area thereafter reduces monotonously. Since $a^T a < 1.0 \times 10^{-6}$ at E (1582 steps), it means that E is a new minimal surface, whose area is 3.50707.

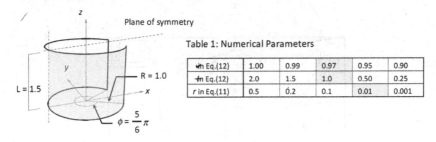

Table 1: Numerical Parameters

	1.00	0.99	0.97	0.95	0.90
α in Eq.(12)	1.00	0.99	0.97	0.95	0.90
λ in Eq.(12)	2.0	1.5	1.0	0.50	0.25
r in Eq.(11)	0.5	0.2	0.1	0.01	0.001

Figure 11.4 Boundary shape.

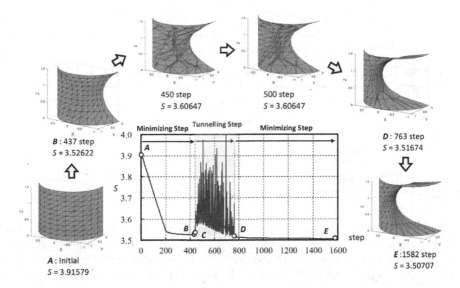

450 step
S = 3.60647

500 step
S = 3.60647

B : 437 step
S = 3.52622

Minimizing Step | Tunnelling Step | Minimizing Step

D : 763 step
S = 3.51674

A : Initial
S = 3.91579

E :1582 step
S = 3.50707

Figure 11.5 The history of area of membrane ($\alpha = 0.97, \lambda = 1.0, r = 0.01$).

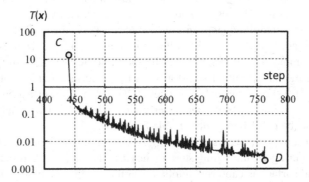

Figure 11.6 Tunnelling function.

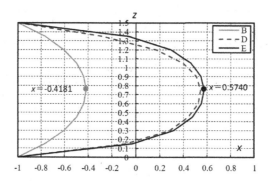

Figure 11.7 Surface section (y = 0).

Table 11.1 The x-coordinate of saddle point on minimal surfaces and the area

	Point B		Point E	
	x-coordinate	Area	x-coordinate	Area
Hinata et al.	−0.3900(C)	3.53272(C)	0.6117(C)	3.50847(C)
	−0.3946(F)	3.53310(F)	0.6036(F)	3.51284(F)
Present	−0.4181	3.52622	0.5740	3.50707

Figure 11.7 shows the section in the symmetry plane ($y = 0$) at B, D, and E in Figure 11.5. Table 11.1 denotes the x-coordinate of the saddle point ($z = 0.75$) on the minimal surfaces at B, E. Compared with the results by Hinata *et al.* (1974), the saddle points are slightly located on the negative side. Symbols (C) and (F) in Table 11.1 mean the results by cylindrical coordinate system and free boundary program, respectively. Pathing through the tunnelling step, the minimal surface, whose area is smaller than the first one, can be confirmed at E. The surface areas at B and E correspond to the results by Hinata *et al.* (1974).

Based on $\alpha = 0.97$, $\lambda = 1.0$ and $r = 0.01$ as numerical parameters, we carry out parametric studies. Figure 11.8 gives the results concerning α. The parameter α means the rate to target surface area during tunnelling step. In the case of $\alpha = 0.95$, the surface which satisfies Eq. (7) cannot be found. For $\alpha = 0.99$, the solution is found for Eq. (7) to finish the tunnel step. Nevertheless, minimal area surface cannot be found in the later minimizing step. The area is 3.52244. During the tunnelling step for $\alpha = 1.0$, the surface remains at the first minimal surface. Figure 11.9 shows the surface shapes at 2000 step. The surface in $\alpha = 0.95$ does not converge to any minimal surfaces, and divergence is observed. This result is different from successful results by $\alpha = 1.0$ to multivariable functions.

Figure 11.10 shows the results concerning λ. The parameter λ means the intensity of the pole in Eq. (12) during the tunnelling step. In the case of $\lambda = 0.5 < 1.0$, as the consequence that the first minimal surface is captured, another minimal surface cannot be found. For $\lambda > 1.5$, the results in Figures 11.10 and 11.11 show that the tunnelling step does not close.

Figure 11.12 shows the results concerning r. The parameter r in Eq. (11) means the norm of the coordinate vector to shift each node at the beginning of the tunnelling step. In

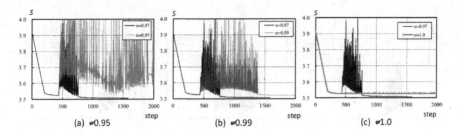

Figure 11.8 The history of area of membrane ($\alpha = 0.95 \sim 1.0, \lambda = 1.0, r = 0.01$).

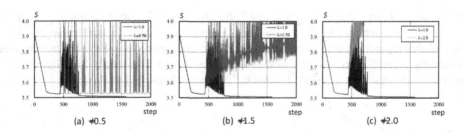

(a) 0.95　　　　(b) 0.99　　　　(c) 1.0

Figure 11.9 Surface shape at 2000 step.

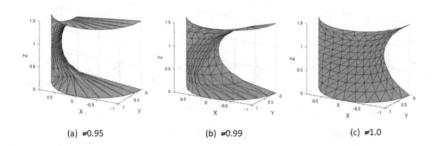

(a) 0.5　　　　(b) 1.5　　　　(c) 2.0

Figure 11.10 The history of area of membrane ($\alpha = 0.97, \lambda = 0.5 \sim 2.0, r = 0.01$).

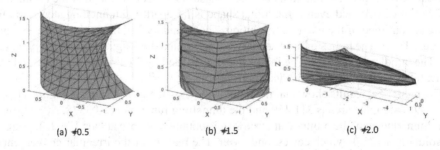

(a) 0.5　　　　(b) 1.5　　　　(c) 2.0

Figure 11.11 Shape at 2000 step.

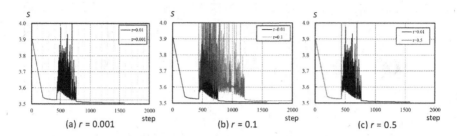

Figure 11.12 The history of area of membrane ($\alpha = 0.97$, $\lambda = 1.0$, $r = 0.001 \sim 0.5$).

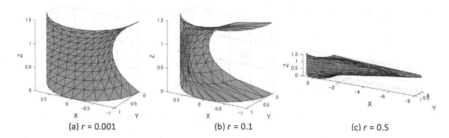

Figure 11.13 Shape at 2000 step.

the case of $r = 0.001$, the norm is too small and insufficient in quantity to search from the first minimal surface. For $r = 0.1$, by way of the tunnelling step, the surface approaches a new minimal surface. In the converged surface, triangular elements near longitudinal edges are densely located in the vicinity of the vertical boundary. The area is 3.51502, which is larger than the least area 3.50707. For $r = 0.5$, the norm is too large to search from the first minimal surface.

3.2 Catenoid

Catenoid is a minimal surface which is spanned between parallel circles as the fixed boundaries. When the distance between boundaries is less than 1.325487 times the radius of the boundary circle, there are minimal and maximal surfaces. Confirming that the minimal area surface can be obtained even if the initial shape is far apart, the tunnelling step starts from the neighbourhood of the maximum solution. The boundary and the initial shape are shown in Figure 11.14. The radius of the boundary circle is 5 and the length between the boundaries is 6. The initial area is 177.8. The initial shape is divided by 24 circumferentially and by 10 vertically using triangular mesh. The control parameters are $\alpha = 0.97$, $\lambda = 1.0$, $r = 1.0$.

The numerical results show that the tunnel step finished in 164 steps in Figure 11.15. At disturbed 1 step, the area is 311.459 and the tunnelling function is 1.03578. At 300 steps in the minimizing step, the minimal area surface is obtained, whose area is 176.2. The analytical solution is 175.0, which corresponds well. The reason for the irregular convergence of the elements in the final convergent shape is considered to be that the element layout at the end of the tunnel step is disordered. Figure 11.16 shows the tunnelling function in Eq. (12) during tunnelling step (1–164 step). The initial value is 1.03578. As in the former example, the tunnelling function shows an inclination to decrease asymptotically. The slight fluctuation of the function is also confirmed. Variation of $T(x)$ is similar to that of the area in Figure 11.15.

As mentioned earlier, the minimal surface search method using the tunnel method was examined, and it was shown that the minimal surface of the smallest area can be obtained by setting appropriate analysis control parameters.

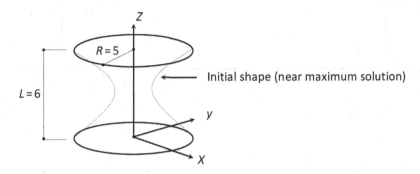

Figure 11.14 Boundary and initial shape.

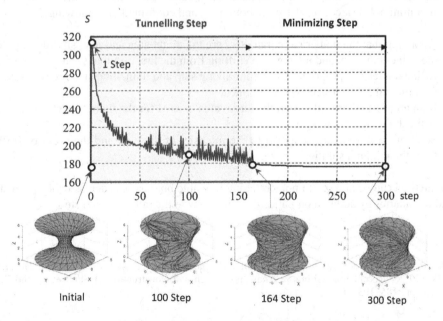

Figure 11.15 The history of area of membrane in Catenoid ($\alpha = 0.97, \lambda = 1.0, r = 1.0$).

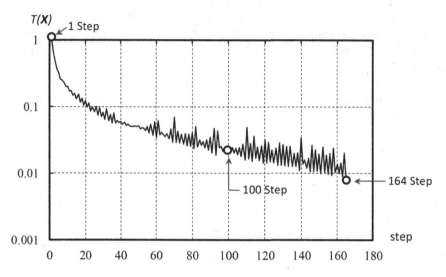

Figure 11.16 Tunnelling function.

4   Summary

A numerical analysis method was presented to search minimal area surface applying the tunnel method, which is a deterministic optimization method of multimodal function. The chapter performed the numerical analysis of the Courant problem and Catenoid that spans different minimal surfaces for the same boundary, and confirmed the following:

1. Assuming that the numerical parameters are the target area and intensity of the pole of the tunnel function, and the norm of shifting from the first minimal solution, and setting each parameter appropriately, the minimizing step and tunnelling step are repeated and the minimal area surface can be obtained.
2. If an area slightly smaller than the target area is used for the tunnel function, it is easy to finish the tunnelling step.
3. Solution search is very sensitive to the intensity of the pole and the norm of shifting from the minimal solution.

This method can be applied to nonstretching movable boundaries such as cables as well as fixed boundaries by adding compatibility conditions in the minimization step.

Note

1 Based on the *Proceedings of the 12th Asian Pacific Conference on Shell & Spatial Structures, APCS2018*, "Recent Innovations in Analysis, Design and Construction of Shell & Spatial Structures," pp. 185–193, 2018.

References

Courant, R. (1950) *Dirichlet's Principle, Conformal Mapping, and Minimal Surfaces.* Interscience, New York.

Frei Otto (1969) *Tensile Structures.* MIT Press, Cambridge, MA and London.

Hildebrant, S. & Tromba, A. (1985) *Mathematics and Optimal Form.* W. H. Freeman & Company, New York, NY and Oxford.

Hinata, M., Shimasaki, M. & Kiyono, T. (1974) Numerical solution of plateau's problem by a finite element method. *Mathematics of Computation*, 28, 125, 45–60.

Ishihara, K. & Ohmori, H. (1993) Shape finding method by using minimal surface. In: SEIKEN IASS, Symposium. (ed.) *Proceedings of Nonlinear Analysis and Design for Shell and Space Structure.* pp. 261–266.

Levy, A.V., Montlvo, A., Gomez, A. & Calderon, A. (1982) *Lecture Notes in Mathematics*, Hennart, J.P. (ed.), No. 909. Springer, Berlin, Heidelberg. pp. 18–33.

Nishimura, T. & Yamanaka, I. (2014) A method for finding minimal surfaces of membranes using simulated annealing. *Journal of the 37th Symposium on Computer Technology of Information, Systems and Applications (Transactions of Architectural Institute Japan)*, pp. 85–90. (in Japanese).

Tsutiya, T. (1986) On two methods for approximating minimal surfaces in parametric form. *Mathematics of Computation*, 46, 174, 517–529.

Chapter 12

Robust design for a large-scale, highly precise space smart structure system

Nozomu Kogiso

1 Introduction

Space missions such as a radio astronomy mission and a satellite communication mission require large aperture reflector systems (Natori *et al.*, 2002; Meguro *et al.*, 2003) because the radiowave with higher frequency has weaker magnitude. In addition, to condense the weak electromagnetic wave to the feeder, the surface shape accuracy is required to be almost 1/10 ~ 1/20 of its wavelength for the radio astronomy mission, or 1/50 ~ 1/100 for the satellite communication. Furthermore, light weight is required for the space structure application as expected.

Such strict design requirements have made the space structure engineering and design progress only so far. As the requirement becomes harder for higher observation frequency, the conventional design methodology is difficult to adopt. For example, the ASTRO-G project, a space radio astronomy VLBI (Very Long Baseline Interferometry) mission that was promoted by Japan Aerospace Exploration Agency (JAXA) but was cancelled in 2011, required 0.4 mmRMS error for the surface shape in 9.4 m aperture reflector with 43 GHz observation frequency (Higuchi *et al.*, 2009). Recently, the balloon-borne VLBI project (Doi *et al.*, 2019) that is planned to observe in about 300 GHz requires 0.05 mmRMS error in 3 m aperture reflector. The conventional structural design approach makes it much more difficult, because the safety margin that considers uncertainty is not sufficient to satisfy the design requirements.

To realize such an antenna system with highly structural design requirements, a new structural design concept is required. In order to satisfy such severe design requirements, several research studies on a smart reflector that can adaptively control the surface shape have been conducted (Fang *et al.*, 2011, Laslandes *et al.*, 2013, Bradford *et al.*, 2012, Datashvili *et al.*, 2010). The purpose of these research studies is to reduce the optical path length error caused by the reflector surface deformation. One promising approach is to adjust the surface shape adaptively.

The author together with several Japanese researchers as a kind of inter-university research project have studied a large-scale highly precise space smart structure system and have proposed a smart Cassegrain antenna system, which is composed of a primary reflector and a smart sub-reflector using piezoelectric actuators (Tanaka *et al.*, 2016, Gotou *et al.*, 2016, Kashiyama *et al.*, 2018). An overview of our proposed prototype is shown in Figure 12.1 (Tanaka *et al.*, 2016). In this system, the distorted wave reflected on the deformed surface of the primary reflector is corrected on the smart sub-reflector that adequately adjusts the path length error, and then the corrected beam is condensed to the feeder. We have proposed that the surface shape of the sub-reflector is properly adjusted using the piezoelectric actuators.

The author has participated on the project team to contribute to the research from the standpoint of structural optimization considering uncertainty, which is the author's major research field. In this chapter, a couple of examples that the structural optimization design is applied to this smart reflector system design problems are introduced.

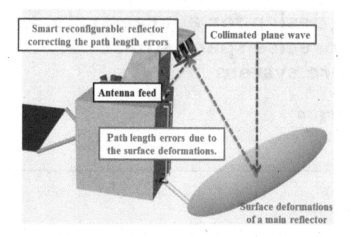

Figure 12.1 Concept of space smart reflector system.

Source: Tanaka *et al.* (2016).

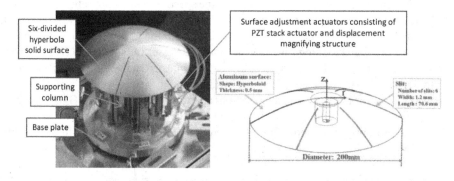

Figure 12.2 Breadboard model of smart reflector and surface adjustment actuator.

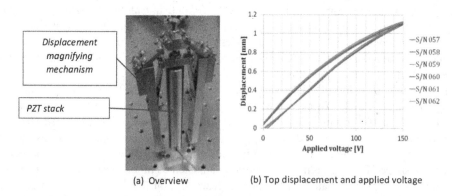

(a) Overview   (b) Top displacement and applied voltage

Figure 12.3 Displacement magnifying mechanism (DMS) with piezoelectric actuator.

2   Proposed smart reflector system

As mentioned in the previous section, the proposed antenna system is composed of a primary reflector and a smart sub-reflector using piezoelectric actuators to adjust the surface shape. The overall picture and the projection view of the breadboard model of the smart sub-reflector are shown in Figure 12.2 (Tanaka *et al.*, 2016). The sub-reflector of 200 mm diameter is divided into six petals and is supported to the base plate by the column located at the centre. An actuator is installed under each petal to give the bending deformation to the petal. The actuator consists of the piezoelectric stack actuator (PSt 150/10/100 VS15; Piezomechanik GmbH) and the displacement magnifying mechanism (DMS) shown in Figure 12.3. Since the stroke of the piezoelectric actuator itself has only 0.1 mm, DMS is installed to magnify the stroke larger than 0.9 mm. The relationship between the top displacement and the applied voltage is shown in Figure 12.3b. Though the generated displacement of the piezoelectric element has hysteresis for loading and unloading and also individual difference between the actuators, the maximum resultant reaction force is confirmed to have more than 90 N that is sufficient to deform the petal of the sub-reflector (Tanaka *et al.*, 2016).

As is the first demonstration's purpose, the smart sub-reflector was installed into the primary reflector of 1.5 m diameter shown in Figure 12.4a that was designed as the prototype of

(a) Photograph of antenna

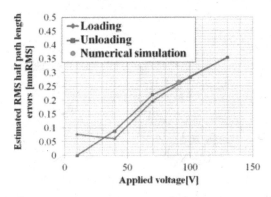

(b) Performance deterioration

Figure 12.4 Prototype of an antenna system with smart sub-reflector.

a balloon-borne VLBI mission (Doi *et al.*, 2019). Because the primary reflector was manu-factured as an ideal parabola shape, the recovery by the sub-reflector actuation was not expected. Instead, the deterioration of the antenna optical performance caused by the surface actuation of the smart sub-reflector was measured by the demonstration experiment and compared with the numerical simulation (Tanaka *et al.*, 2016). Figure 12.4b summarizes the comparison of the estimated half-path length errors by the experiment with loading and unloading paths and the numerical simulation.

Although this demonstration validates the feasibility of the smart sub-reflector, the tech-nical readiness level is still poor to realize this smart sub-reflector. This chapter introduces our research activities to improve the technical readiness level of the smart sub-reflector system from the standpoint of the structural optimum design considering uncertainty.

3   Example of application of structural optimization for the smart reflector problem

3.1   Primary reflector design problem as multi-objective optimization considering uncertainty

As the first example, the primary reflector design formulated as using the multi-objective optimization method is introduced. Even though the smart sub-reflector can adjust the opti-cal path length error, the primary reflector should have sufficient surface accuracy under several nominal load conditions. This study uses the primary reflector for the next-stage bal-loon-borne VLBI mission (Kogiso *et al.*, 2016) that the reflector is planned to have 3-meter aperture consisting of iso-grid structure of aluminium alloy as shown in Figure 12.5a. The skin thickness, the rib thickness, and height, as shown in Figure 12.5b, are treated as design variables. The primary reflector is connected to the backup structure through kinematic cou-plings (KCs), as illustrated in Figure 12.6. The KC positions and the direction angle of the mating V-groove of KC are also treated as design variables.

Several load conditions listed in Table 12.1 are considered corresponding to the self-weight deformation under several elevation angles and the thermal deformation under the temperature difference from the ground to 30,000 m height. The nominal condition is set as 45 degrees of elevation angle and −70 °C of the temperature change. Other combinations of

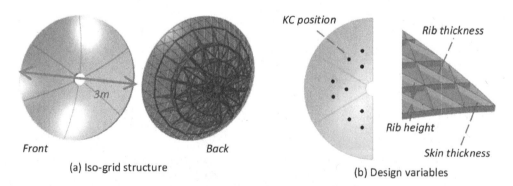

(a) Iso-grid structure

(b) Design variables

Figure 12.5 Primary reflector for next-stage balloon-borne VLBI mission.

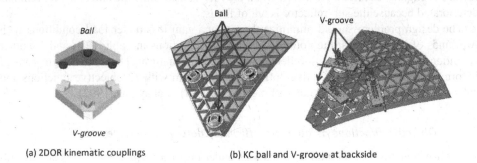

(a) 2DOR kinematic couplings

(b) KC ball and V-groove at backside

Figure 12.6 Kinematic couplings allocated on the backside of primary reflector.

Table 12.1 Load cases

Case	Elevation angle (deg)	ΔT (°C)
1	10	−55
2	10	−85
3	45	−55
4 (nominal)	45	−70
5	45	−85
6	70	−55
7	70	−85

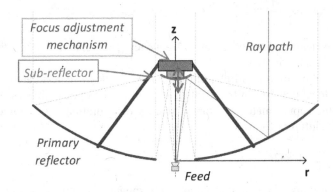

Figure 12.7 Sub-reflector adjustment system.

10 to 70 degree of elevation angles and −55°C to −85°C of the temperature range listed in Table 12.1 are considered as load conditions considering uncertainty.

In addition, the failure of the focus adjustment mechanism of the sub-reflector shown in Figure 12.7 is considered. When the mechanism works, the antenna gain can be maximized by adjusting the focal point by minimizing the optical path length error, even if the primary surface is slightly distorted. However, when the mechanism fails, the antenna gain remains deteriorated because the sub-reflector is out of focus.

The design problem should minimize the antenna gain loss under load conditions with two kinds of uncertainties: the nominal seven load conditions in Table 12.1, and six other conditions with failure of the sub-reflector adjustment mechanism. Then, the design problem is formulated as the multi-objective optimization problem with 13 objective functions and solved by using the sacrificing trade-off method (STOM) (Nakayama, 1992).

3.1.1 Objective functions as aperture efficiency deterioration rate

The self-weight and the thermal deformations under the observatory conditions are considered to deteriorate the surface shape, and the deformations are evaluated by using the nonlinear FEM tool. After obtaining the deformation distribution, the RMS error from the ideal parabola shape is evaluated using the following equation:

$$\varepsilon_{\text{RMS}} = \left(\sum_{i=1}^{N} A_i \varepsilon_i \bigg/ \sum_{i=1}^{N} A_i \right)^{1/2} \tag{1}$$

where N is the number of nodes, ε_i is the surface error at the i-th node and A_i is the electric field intensity using the following simple tapered pattern:

$$A(\theta) = A_0 10^{A_x (\theta/\theta_0)^2 / 20} \tag{2}$$

Then, the aperture efficiency is evaluated from the following Ruze equation that indicates the relationship between the RMS error and the deterioration of the electromagnetic wave intensity receiving at the distorted reflector surface:

$$\eta_R \left(\varepsilon_{\text{RMS}} \right) = \exp \left(-\frac{4\pi \varepsilon_{\text{RMS}}}{\lambda} \right) \tag{3}$$

where λ is a wavelength of the observatory frequency. This relationship is established when the focal position is perfectly set.

When the focal position is shifted, the aperture efficiency is deteriorated. The deterioration due to the focal position shift is evaluated by using the optical physics software. After trial analyses, the approximated aperture efficiency deterioration ε_s is obtained in terms of the focal length shift δ as follows.

$$\eta_s \left(\delta \right) = \exp \left[-\left(1.182\delta \right)^2 \right] \tag{4}$$

The value of η_s is unity when the sub-reflector adjustment mechanism works as $\delta = 0$ and the value becomes larger as the gap δ between the focal and the mechanism position becomes larger.

Then, the objective functions are formulated as follows:

$$\text{Minimize: } f_{\text{normal}} = 1 - \eta_R \left(\varepsilon_{\text{RMS}} \right) \tag{5}$$

$$\text{Minimize: } f_{\text{fail}} = 1 - \eta_s \left(\delta \right) \cdot \eta_R \left(\varepsilon_{\text{RMS}} \right) \tag{6}$$

where Equation (5) corresponds to the seven nominal conditions and Equation (6) to the six other conditions when the sub-reflector adjustment mechanism fails. Here, Case 4 listed in Table 12.1 is set as a nominal condition. That is, the sub-reflector position will be fixed at the best position for this case if the adjustment mechanism fails. It should be noted that Equation (6) corresponds to the worst-case design, one of the robust design formulation.

The design requirement for the aperture efficiency deterioration should be less than 17% for all conditions.

3.1.2   Numerical optimum solutions

An example of the obtained optimum solution that satisfies the design requirement of the aperture efficiency deterioration level less than 17% is found by STOM as listed in Table 12.2. The maximum aperture efficiency deterioration level is 16.6% in Case 1 under failure in the sub-reflector adjustment mechanism. The deformation surface error distributions are illustrated in Figure 12.8. The positive value corresponds to the error to the front direction and the negative

Table 12.2 Aperture efficiency deterioration level (%) for optimum design

Adjustment mechanism	Cases						
	1	2	3	4	5	6	7
Normal	3.06	9.07	2.53	1.82	2.41	6.53	2.45
Failure	16.6	13.5	11.2	1.82	11.1	12.3	14.1

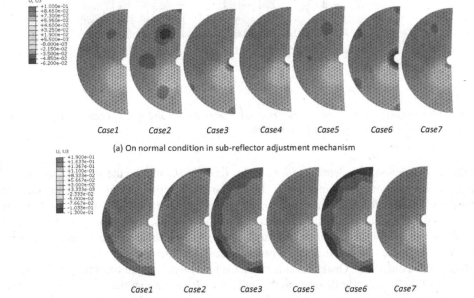

(a) On normal condition in sub-reflector adjustment mechanism

(b) On failure condition in sub-reflector adjustment mechanism

Figure 12.8 Surface error distribution for optimum design.

Table 12.3 Focal position shift and sub-reflector adjustment

Case	1	2	3	4	5	6	7
δ(mm)	1.009	1.525	1.079	1.337	1.594	1.123	1.639
Adjustment (mm)	0.328	−0.188	0.258	–	−0.258	0.213	−0.302

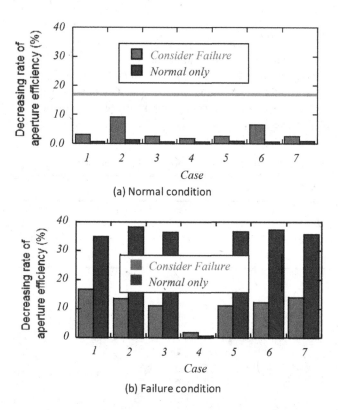

(a) Normal condition

(b) Failure condition

Figure 12.9 Comparison of aperture efficiency deterioration with and without considering failure on sub-reflector adjustment mechanism.

to the back direction. The deformation range is −0.062 to 0.1 mm on the normal condition. A large surface error occurs around the KC position in Case 2, that has the maximum aperture efficiency deterioration rate in normal condition as less than the threshold value. However, the range is −0.13 to 0.18 mm on the failure condition that is almost twice as that on the normal condition. On the failure condition, the large error is significant around the outer circumference region. In the higher temperature cases (Cases 1, 3, and 6; $\Delta T = -55°C$), the reflector deforms to the outer side. On the other hand, in the lower temperature cases (Cases 2, 5 and 7: $\Delta T = -85°C$), the reflector deforms to the inner side. It means that the thermal deformation has a significant effect on the aperture efficiency deterioration. This tendency is found from the focal position shift listed in Table 12.3. The adjustment length is positive for Cases 1, 3, and 6, but negative for Cases 2, 5, and 7.

Finally, Figure 12.9 compares the optimum design considering the failure on adjustment mechanism with the design, without considering the failure condition. When only the nominal conditions are considered, the optimum design is superior in the nominal conditions to the design considering failure conditions. However, if the adjustment mechanism fails, the optimum design without considering failure conditions deteriorates the performance significantly beyond the requirement level. On the other hand, the optimum design considering the failure conditions satisfies the design requirement.

3.2 Optimum design of displacement magnifying mechanism as a compliant mechanism

The prototype of DMS shown in Figure 12.3 has a lot of fastening parts that cause friction loss or dimension instability. To reduce the looseness, DMS is considered to produce as a compliant mechanism that is known as a jointless single-piece structure to utilize the elastic deformation for transferring an input displacement to an output displacement to the other position.

The structural shape is obtained through formulating the design problem as structural optimization. The DMS shape of the initial design and the design variables are illustrated in Figure 12.10a. The objective function is to maximize the tip displacement that is transferred from the actuation displacement of 0.08 mm. The constraints are the maximum allowable stress and the first eigenfrequency limit. The other design limitations as buckling load and the tip reaction force are not included in the constraint condition after some preliminary optimization trial.

The prototype of the optimum design produced by a wire electrical discharging machining is shown in Figure 12.10b. When the input displacement to the upper direction is given by the actuator, the leftmost and the rightmost points A and B are deformed to the narrowing

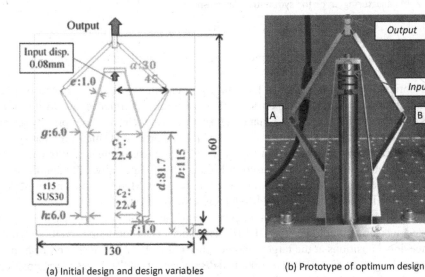

(a) Initial design and design variables   (b) Prototype of optimum design

Figure 12.10 Displacement magnifying mechanism design as compliant mechanism.

(a) Image of 3D shape measurement device   (b) Measurement camera and DMS on glass table

Figure 12.11 Precise 3D shape measurement device.

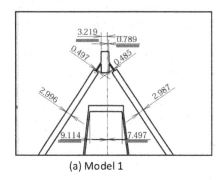
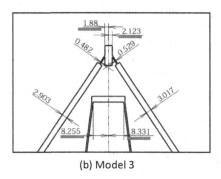

(a) Model 1   (b) Model 3

Figure 12.12 Manufacturing error for symmetric direction.

direction and then the top point is deformed to the upper direction almost 10 times as much as the input deformation.

As the optimum design is found to have thin portions that will act as a kind of plate springs, the effect of manufacturing error on the structural performance should be investigated (Kogiso *et al.*, 2017). For the manufacturing accuracy check, three prototypes are produced, and then the dimensions are measured by using a precise 3D measurement system (Mitsutoyo Quick Vision Pro) shown in Figure 12.11. The measurement results with most and least asymmetric models are compared in Figure 12.12. Especially, the manufacturing error to the symmetric direction is found on the top position. Model 1, shown on the left-hand side, has the largest difference and Model 3 has the smallest.

Then, the displacement of DMS when the piezoelectric actuator works is measured by the stereo image measuring method called DLT (Direct Linear Transformation) method that the three-dimensional positions at the target markers put on the object are measured from multiple cameras (Kogiso *et al.*, 2017). The obtained displacements at the representing positions are compared with those by FEM analysis where the geometric models are constructed from the shape measurement result. As indicated in Table 12.4, Model 1 has larger declination to

Table 12.4 Comparison of displacement between experiment and simulation

Displacement (mm)	Model 1		Model 3	
	Simulation	Experiment	Simulation	Experiment
Input	0.083	0.095	0.084	0.093
Output to upper	0.997	1.09	0.979	1.046
Output to side	0.340	0.103	0.039	0.059
Point A side	0.820	1.090	1.051	1.059
Point B side	1.298	1.293	1.005	1.147

the horizontal direction. On the other hand, Model 3 has a smaller declination to the horizontal direction.

The individual difference of dimensions between DMSs will have a significant effect on the adjusting performance of the smart sub-reflector. On the other hand, strict geometric tolerance makes the cost more expensive. Therefore, further study to investigate the effect of the geometric deviation is required.

4   Summary

This chapter introduces our project research concerning the large-scale highly precise space smart structure. Then, the design examples applying the structural optimization method considering uncertainty were illustrated. One was the primary reflector design problem of the iso-grid structure. The design problem was formulated as a multi-objective optimization problem considering several load conditions and the effect of adjusting mechanism failure. The other was the displacement magnifying structure design constructed as the compliant mechanism. Then, the effect of manufacturing error was investigated.

Acknowledgements

This research is partially supported by Strategic Research Grant of Institute of Space and Astronautics Sciences of JAXA (ISAS/JAXA) and JSPS KAKENHI 26249131. The author appreciates colleagues of this research project; Prof. Hiroaki Tanaka (National Defense Academy of Japan), Prof. Takashi Iwasa (Tottori University), Prof. Takeshi Akita (Chiba Institute of Technology), Prof. Hiraku Sakamoto (Tokyo Institute of Technology), Prof. Tadashige Ikeda (Chubu University), Prof. Atsuhiko Senba (Meijo University), and Prof. Kosei Ishimura (Waseda University), who have given me the wonderful opportunity to research together.

References

Bradford, S.C., Agnes, G.S., Ohara, C.M., Shi, F., Peterson, L.D., Hoffman, S.M. & Wilkie, W.K. (2012) Piezocomposite actuator arrays for correcting and controlling wavefront error in reflectors. *53rd AIAA/ASME/ASCE/AHS/ASC Structures, Structural Dynamics and Materials Conference, 23–26 April, Honolulu, Hawaii, USA,* AIAA 2012–1743.

Datashvili, L., Baier, H., Wei, B., Hoffman, J., Wehrle, E., Schreider, L., Mangenot, C., Santiago-Prowald, J., Scolamiero, L. & Angevain, J.C. (2010) Mechanical investigations of in-space-reconfigurable reflecting surfaces. *32nd ESA Antenna Workshop on Antennas for Space Applications, 5–8 October, Noordwijk, Netherland*, pp. 1–8.

Doi, A., et al. (2019) A balloon-borne very long baseline interferometry experiment in the stratosphere: Systems design and developments. *Advances in Space Research*, 63(1), 779–793.

Fang, H., Quijano, U., Bach, V., Hill, J. & Wang, K.W. (2011) Experimental study of a membrane antenna surface adaptive control system. *52nd AIAA/ASME/ASCE/AHS/ASC Structures, Structural Dynamics and Materials Conference, 4–7 April, Denver, Colorado, USA*, AIAA-2011–1828.

Gotou, K., Sakamoto, H., Inagaki, A., Tanaka, H., Ishimura, K. & Okuma, M. (2016) Actuator design for space smart reflector to reduce thermal distortion. *Transactions of the Japan Society for Aeronautical and Space Sciences, Aerospace Technology Japan*, 14(30), Pc_25–Pc_31.

Higuchi, K., Kishimoto, N., Meguro, A., Tanaka, H., Yoshihara, M. & Iikura, S. (2009) Structure of high precision large deployable reflector for space VLBI (very long baseline interferometry). *50th AIAA/ASME/ASCE/AHS/ASC Structures, Structural Dynamics and Materials Conference, 4–7 May, Palm Spring, California, USA*, AIAA-2009–2609.

Kashiyama, R., Sakamoto, H., Okuma, M., Tanaka, H. & Ishimura, K. (2018) Athermalization of deformable reflector's actuators for radio astronomy satellites. *AIAA SciTech 2018, 8–12 January, Kissimmee, Florida, USA*, AIAA-2018–1199.

Kogiso, N., Kodama, R., Kimura, K., Satou, Y., Doi, A. & Tanaka, H. (2016) Balloon-borne VLBI reflector structural design considering failure of sub-reflector adjustment mechanism using multiobjective optimization. *Aerospace Technology Japan*, 15, 91–100. (in Japanese).

Kogiso, N., Furutani, N., Naka, T., Kimura, K, Tanaka, H. & Iwasa, T. (2017) Optimum structural design for high-precision space smart reflector. *12th World Congress on Structural and Multidisciplinary Optimization, 5–6 June 2017, Branschweig, Germany*, ID: 261.

Laslandes, M., Hugot, E., Ferrari, M., Hourtoule, C., Singer, C., Devilliers, C., Lopez, C. & Chazallet, F. (2013) Mirror actively deformed and regulated for applications in space: Design and performance. *Optical Engineering*, 52(9), 091803.

Meguro, A.m Harada, S. & Watanabe, M. (2003) Key technologies for high-accuracy large mesh antenna reflectors. *Acta Astronautica*, 53(11), 899–908.

Nakayama, H. (1992) Trade-off analysis using parametric optimization techniques. *European Journal of Operational Research*, 60(1), 87–98.

Natori, M.C., Hirabayashi, H., Okuizumi, N., Iikura, S. & Nakamura, K. (2002) A structure concept of high precision mesh antenna for space VLBI observation. *43rd AIAA/ASME/ASCE/AHS/ASC Structures, Structural Dynamics and Materials Conference, 22–25 April, Denver, Colorado, USA*, AIAA-2002–1359.

Tanaka, H., Sakamoto, H., Inagaki, A., Ishimura, K., Doi, A., Kono, Y., Oyama, T., Watanabe, K., Oikawa, Y. & Kuratomi, T. (2016) Development of a smart reconfigurable reflector prototype for an extremely high-frequency antenna. *Journal of Intelligent Material Systems and Structures*, 27, 764–773.

Chapter 13

Instruction for beginners in studying design

Sun-Woo Park

1 Introduction

I remember the phrase "One of the engineer's main roles is to help architects design," which was often said by my thesis director, a structural engineer, when I studied in Germany.

In structural planning, two approaches can be used: numerical analysis and consultation during the design process. Therefore, from the beginning, it is desirable for the stakeholders to have a consultation process to derive a consensus on overall structural stability without the attitude of considering design and structure as a parallel relationship.

First, the determination of the member size is not a priority in the basic design process. Rather, it is important to propose a reasonable structural system to the architect considering the overall stability first, and then to determine the final member size based on the completed design drawings. In other words, during the basic design process, it may be more effective to observe the working model and exchange ideas rather than to perform numerical analysis and determine the size of the structural component.

In the following section, the efficiency of the design process using the modelling mentioned here will be described through some of my successful examples.

2 Design examples using modelling

Although many model works have been carried out so far, the example introduced here will be limited to the scope of dome structure and footbridge.

2.1 Domes

There are several different modes of driving the roof. It can be an event for an audience just by watching a moving figure. Generally, when a person watches a moving object, it is said that the attention decreases after 20 minutes. For this reason, it is common for most dome roofs to be planned to run within 20 minutes.

In general, for dome structures composed of spatial trusses, membranes, ribs, and arches, stakeholders can easily communicate using sketches. However, in the case of cable or retractable domes and special structures, it is much more efficient to collaborate in the design process using real models rather than sketches.

2.1.1 Tensegrity domes (Park, 2015a)

Understanding the structural system through general sketches is not an easy problem. It is also a difficult problem to understand how to build the system and the process. These problems were clearly understood in the process of actually building models.

Figure 13.1 Working process.

Figure 13.2 Completed model.

The models shown in Figures 13.1 and 13.2 were manufactured to the maximum size as possible using commercially available MDF. It was a challenge to keep the flight mast vertical during the modelling.

The first method used compression and tensile ring. For this, the compression ring was cut with a diameter of 120.0 cm, and two-layer tension rings were fabricated in the centre. On these three rings, a compression rod was inserted vertically through the holes drilled at the planned points. In this state, the compression rod was vertically erected by pulling the tensile materials. However, when the three rings were removed, the balance of force collapsed and the compression rod could not be held vertically.

As a result of applying several methods, the most appropriate method was to balance the force by adjusting the tension, by slowly pulling the tensile material at each anchorage point. The building construction process was also achieved by gradually tensioning to each cable at anchorage points on the compression ring after completing the internal member configuration, as similar to the conclusion obtained through modelling.

2.1.2   Rotation dome (Park, 2015a)

The most important thing of the retractable dome is the driving method of the roof. When surveying the retractable domes built in various parts of the world, these domes can be classified according to the driving method in general. The driving method can also be understood through modelling. The driving method of the dome shown in Figures 13.3 and 13.4 has not yet been realized.

The overall shape of the dome is partitioned in such a way that the lower part is fixed and only the upper part is opened and closed. The basic concept of this retractable system is that two divided roofs move while rotating in the opposite direction. Four sensors are planned to be installed and controlled for the actual opening and closing process. This process is executed automatically when the operation button is pressed. The general public will be able to easily understand how to operate this retractable dome by just looking at how the model works.

Figure 13.3 Study model of the driving method.

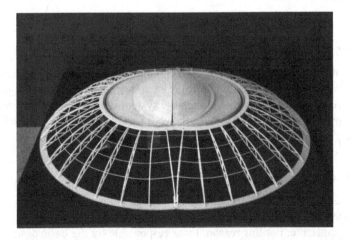

Figure 13.4 Completed model.

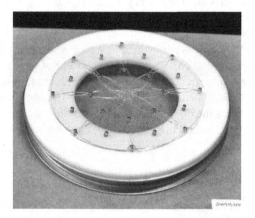

Figure 13.5 Study model of the driving method.

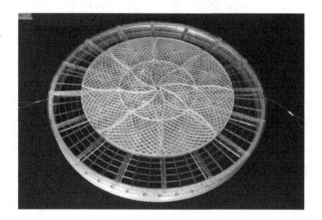

Figure 13.6 Completed model.

2.1.3 Flower dome (Park, 2015a)

A scene while the dome is opened and closed is reminiscent of the shape of a flower blooming, or the movement of the filter to control the exposure of the camera (Figures 13.5 and 13.6). The mode of operation for this retractable system is unique, but a substantial amount of power energy will be required. Because the roof surfaces move on a plane without using the counterpart weight, there is a disadvantage as a significant weight act.

2.2 Footbridge

Of the two areas of architectural structure and civil engineering, who is in charge of the design of the bridge depends on the circumstances of each country. In Korea, it is mainly civil engineering, not architecture. For this reason, I was responsible for the basic design and concept, and the final structural analysis and construction drawings were completed by civil engineers.

Figure 13.7 Study model.

Figure 13.8 Photo of the actual bridge.

2.2.1 TEDA footbridge, China (Park, 2015b)

The bridge was actually constructed and is located in the TEDA/Tianjin Economic Development Area within TEDA Harbor City in suburb of Tianjin, China (Figures 13.7 and 13.8).

Provided a variety of four basic plans, the client chose the concept. The basic concept of the selected proposal is to depict vertical components installed on the sailboat of a vessel anchored at the dock. It is a cable-stayed bridge with a span of about 120.0 m, and the deck is planned in a straight line. The pylon is installed with a 45-degree incline over the deck, and the deck of the bridge is suspended by cables from the bent portion of the pylon. The cables arranged in two directions on the rear of the pylon were anchored on pedestals in each direction in order to balance and support the forces acting in this state.

I provided the basic design and concept, and the construction drawings were completed by a Chinese construction company.

2.2.2 Dabong footbridge, Korea (Park, 2015a)

This bridge of the project that won the contest was scheduled to be constructed at the entrance to Seoul but just finished with only a plan (Figures 13.9 and 13.10).

The basic concept of planned features is to remind people of the scenery of surrounding mountains. It is a cable-stayed bridge using circular members with different diameters as

Figure 13.9 Study model.

Figure 13.10 Rendering.

pylons. Two circular pylons with different diameters are connected to each other using tension bars, and the bridge decks are suspended from cables at both sides of the pylons. Cables at the other sides were anchored on the pedestrian deck. By placing the cables from the high point of the circular pylon to the nearest horizontal point on the deck, the closer the cable is located to the bottom of the pylon, the farther the anchorage point on the deck. That is, the same as the method of producing HP shells using straight-line materials. From a side view, it is designed to form a curve similar to the aspect of a mountain.

2.2.3 Ssetgang footbridge, Korea (Park, 2015a, 2015b)

This bridge, which won the project contest, connects Shingil-dong and Yeouido across the Han River. It is an asymmetric cable-stayed bridge with two tilted pylons, and the bridge deck is an S-shape derived from the shape of the Han River flowing beneath (Figures 13.11 and 13.12).

It is the largest pedestrian bridge in Korea, with a total length of 320 m and a span between pylons of 200 m. The bridge was initially planned as an asymmetric suspension bridge but was later changed to a cable-stayed bridge due to a technical difficulty. The lower the point where the cable is connected to the pylon, the farther the anchorage point on the

Figure 13.11 Study model.

Figure 13.12 Photo of the actual bridge.

deck, so that the side view is curved. This is also the same as the method of creating HP-shell using straight-line materials. Its outline viewed from the side is a shape of catenary, and from the riverside it is perceived as a huge V-shape. The upper part of pylons including the deck is a steel structure, while the lower part is made of RC.

3 Conclusion

The effectiveness of modelling work has been demonstrated through the lessons with students and the projects actually involved. The basic philosophy of the structural planning I felt through such those processes can be summarized as follows.

Basic principles of structural system planning

- Beautiful and natural structure
  What is a desirable structure? It will not be easy for everyone to define. However, it would be the best goal for a structural engineer to sit face to face with an architect and pursue each other's beauty.

- Simple flow of force
  The longer the flow of forces, the more components it takes. In other words, it becomes an uneconomical structure. A concise structural plan is required by simplifying the flow of forces as much as possible.

- As the water flows
  The final goal of the structure we think is to transfer all the loads acting on the building to the ground. The rain naturally flows down into the ground. The flow of force should also be planned on the same principle as the flow of water.

- Prevention of the domino phenomenon of collapse
  One of the main goals of structural engineers is to ensure the safety of users. Especially in large-scale, large-space structures, structural plans should be performed so that the domino phenomenon of collapse does not occur. In other words, even if a part of a building collapses, the other part should be safe so that the users can be evacuated.

- Maximization of cross-sectional performance of materials
  The cross-sectional size of structural members depends largely on the bending moment. It is effective to plan the structure so as to bear the bending moment as small as possible, and to bear mainly compressive and tensile forces like arch structure and suspension structure.

References

Sun-Woo Park (2015a) *Essay of Architectural Structure*. Pub. Seoul, Korea: Woori Book.
Sun-Woo Park (2015b) *A Miscellaneous Talking of Architectural Structure*. Pub. Seoul, Korea: Woori Book.

Printed in the United States
by Baker & Taylor Publisher Services